HEALTH AND DEVELOPMENT

Health is an elusive concept, and environmental change is leading us to new health risks and exposure to new forms of danger. Development, in its many facets, can have very positive but also negative effects on physical and mental health. Poverty continues to afflict Third World cities and increasing numbers in the developed world, while macroeconomic policies, recession in the West, debt servicing in the Third World and the impacts of structural adjustment policies often mean cutbacks in already poor health services and increases in differential access to resources and health care. Yet people are living longer, their lifestyles are changing and so too are the diseases from which they suffer.

Clearly structured, with regional and country case studies accompanying conceptual and thematic sections, *Health and Development* discusses the multifaceted aspects of health impacts of development. It focuses on crucial issues such as the effects of economic adjustment and environmental change on health, the possibility of extending health services, socio-cultural factors in HIV/AIDS transmission, and the health of women beyond maternal and child health.

If development is to be associated with longer, healthy life rather than a deteriorating quality of life, then the issues raised in this book must be addressed by policy-makers, medical and paramedical staff and researchers worldwide.

David R. Phillips is Director of the Institute of Population Studies and Reader in Health Studies at the University of Exeter, and Chair of the International Geographical Union Commission on Health, Environment and Development. **Yola Verhasselt** is Professor of Geography at the Free University, Brussels, and was Chair of the International Geographical Union Commission on Health and Development.

HEALTH AND DEVELOPMENT

Edited by David R. Phillips and Yola Verhasselt

London and New York

First published 1994

Reprinted 2001 by Routledge
11 New Fetter Lane
London EC4P 4EE

Simultaneously published in the USA and Canada
by Routledge
29 West 35th Street, New York, NY 10001

Reprinted 1998

Routledge is an imprint of the Taylor & Francis Group

British Library Cataloguing in Publication Data
A catalogue record for this book is available from the British Library

Library of Congress Cataloging in Publication Data
A catalog record for this book is available from the Library of
Congress

ISBN 0–415–08528–4 (hbk)
ISBN 0–415–08529–2 (pbk)
Transferred to digital reprinting 2001
Printed in Great Britain by Antony Rowe Ltd, Eastbourne

CONTENTS

CONTENTS

vi

CONTENTS

FIGURES

TABLES

CONTRIBUTORS

Professor Rais Akhtar – Department of Geography and Regional Development, University of Kashmir, Srinagar, India.

Dr Sheena Asthana – Department of Geography, University of Exeter, UK.

Graham Bentham – School of Environmental Sciences, University of East Anglia, Norwich, UK.

Dr Susana Isabel Curto de Casas – Sociedad Argentina de Estudios Geográficos, Buenos Aires, Argentina.

Professor Fang Ru-Kang – Department of Geography, East China Normal University, Shanghai, People's Republic of China.

Dr Nicholas Ford – Institute of Population Studies, University of Exeter, UK.

Dr Wil Gesler – Department of Geography, University of North Carolina at Chapel Hill, USA.

Dr Trudy Harpham – Urban Health Programme, London School of Hygiene and Tropical Medicine, UK.

Dr B. Hyma – Department of Geography, University of Waterloo, Ontario, Canada.

Dr Bose F. Iyun – Department of Geography, University of Ibadan, Nigeria.

Dr Nilofar Izhar – Department of Geography, AMU, Aligarh, India.

Professor Alun E. Joseph – Department of Geography, University of Guelph, Ontario, Canada.

Dr Edith C. Kieffer – Department of Public Health, Michigan State University, Ann Arbor, USA.

Dr Helmut Kloos – Department of Geography, Addis Ababa University, Ethiopia.

Dr Nancy Davis Lewis – Department of Geography, University of Hawaii at Manoa, Honolulu, Hawaii, USA.

Professor Anne Martin-Matthews – Department of Family Studies, Gerontology Research Centre, University of Guelph, Ontario, Canada.

Dr Eva Orosz – Department of Social Policy, Eötvös Loránd University, Budapest, Hungary.

Professor Cosimo Palagiano – Istituto di Geografia, Università degli Studi di Roma, Città Universitaria, Rome, Italy.

Dr David R. Phillips – Institute of Population Studies, University of Exeter, UK.

Penny Price – Department of Geography, University of Exeter, UK.

Professor A. Ramesh – Department of Geography, University of Madras, India.

Professor Yola Verhasselt – The Free University, Brussels, Belgium.

Professor Anthony M. Warnes – Department of Geography, King's College, London, and Age Concern Institute of Gerontology, UK.

PREFACE

The relationships and interactions between health and development are complex. It is by no means clear that health status automatically improves with rising levels of development in any given country, and this certainly cannot be said for all inhabitants. Both the concepts 'health' and 'development' are, in any case, notoriously difficult to define and almost impossible to quantify in a way that all would find acceptable. Nevertheless, health care professionals, researchers and policy-makers are all aware that health status *is* changing with development, and it is not invariably changing for the better. Many people are living longer but there is often a question mark over the quality of life years added, and there are large numbers of people in many countries for whom development has not led to health improvements. For example, many people, especially among the poor and particularly in the so-called Third World, are experiencing twin threats from infectious and chronic/degenerative diseases. These problems are not confined solely to poor countries, and retrenchment in public-sector and private expenditure on health and health care is now commonplace, meaning that publicly provided health care is often diminishing and overstretched.

It was against this background that a concerted effort to study the spatial and social aspects of health and development was undertaken under the aegis of the International Geographical Union. A Commission on Health and Development was established in 1988 under the chairmanship of Professor Yola Verhasselt, with David Phillips as its Secretary. The initial life span of the commission was until 1992, when the remit of the commission was extended to include health, *environment* and development, and David Phillips became its Chair. The intervening years had brought to the fore the influence that environmental change was having and was likely to have on human health at global and local scales. The majority of the members of the original commission have contributed to this book on health and development, which has come about largely as a result of the concentrated research and international collaboration and exchanges engendered by the IGU commission. Other experts in their fields have been invited to contribute chapters on specific issues, so that a broad coverage of health and development can be given. We intend that this book should act both as a research resource for people already working in the development and health care fields and also as a comprehensive source text for higher-level undergraduates and researchers in universities studying development studies, geography, social policy and courses allied to social medicine.

The book is structured in three main parts, intended to progress logically. Part I comprises six scene-setting chapters which introduce conceptual and background issues in health and development. After an initial chapter outlining major issues such as health transition, health and poverty, and the other main concerns of the book, chapters focus on the impacts on health of global environmental change and of economic/structural adjustment policies being pursued in many countries at present. Chapters then discuss the potential for traditional medicine to contribute to health care; the crucial matter of socio-cultural and development factors underlying the global transmission of a major threat to health and development, HIV/AIDS; and the sometimes dubious relations between the pharmaceutical industry and health, particularly in some Third World countries.

Part II also consists of six chapters. In this part, the intention is to discuss the health care needs of specific groups and persons and the impacts on them of various aspects of development. A chapter on the health of urban residents, particularly focusing on the urban poor, is followed by two chapters which look at maternal and child health policies and the important subject of women's health, which has often to date been subordinated in policy and practice to the health needs of children. Two chapters focus on the demographic, health care and economic consequences, all highly inter-related, of the ageing of populations. The first of these chapters considers the ageing of populations in the Third World, where many policies are still geared to dealing with infectious conditions, the health of children and family planning or reproductive health. The longer-term consequences of population ageing are highlighted by the second chapter, which discusses the workforce and caring issues raised by the growing numbers of elderly people in Canada, extrapolating from there to the future prospects for Third World countries in which pension and social care systems are currently much less developed. A final chapter in this section reviews the key role of comprehensive and selective primary health care policies and the potential and limitations of community participation in sustaining health care in countries where resources are severely limited.

Part III of the book provides regional and country-specific studies of health and health care changes with development. Several chapters focus on the problems of Third World countries, with a particular emphasis on poorer countries. Many authors make the point that it is unwise to try to understand contemporary health care patterns and problems without reference to the history of specific countries. In particular, the grafting of modern health care systems on to traditional systems and the selective rationing of modern care under many colonial regimes have imparted a particular character to the health systems in many countries in Africa, South Asia and Latin America. By contrast, erstwhile allegedly equal and modern health care delivery systems in (former) state-socialist countries are today often under great stress and are sometimes unable to provide basic care for many citizens. Wholesale restructuring of the financial and reward elements are probably needed before these health care systems can again start to meet the health needs of their populations.

In terms of presentation of individual chapters, we have specifically asked authors to try to highlight key trends and to be selective in the references they have cited. To some extent, the reference lists provide a guide to further reading rather than a comprehensive bibliography of often inaccessible literature. In Chapter 1, we have attempted to provide a more comprehensive listing of references of interest to researchers and to provide guidance as to statistical sources available for social research in the health field. We hope that this format will assist readers and signpost the more important areas for further reading. In Chapter 20, we adopt a retrospective view of the major themes and issues highlighted in the book, and we also look forward over the next decade or so, to highlight those which will be likely to increase in importance.

Finally, we wish to acknowledge many organizations and individuals who have helped directly and indirectly in the genesis of this book. The International Geographical Union provided official and some financial support for the preparation of final versions of the text. Fourteen international meetings were held between 1988 and 1992 under the auspices of the IGU Commission on Health and Development. We wish to acknowledge financial support for these meetings from a number of sources, in particular the British Council, the Commission of the European Communities, the Carnegie Foundation and the Commonwealth Foundation. Their support assisted with the organization of these meetings and enabled the participation of scientists from developing countries. It was largely as a result of issues identified and collaboration fostered via these meetings that the book has come about. The British Council, through its support for a link between Exeter University and the Free University of Brussels, was instrumental in enabling the collaboration between the editors of the book. In terms of the production of the book, we wish to acknowledge the secretarial and clerical assistance of the staff of the Institute of Population Studies at the University of Exeter who worked on manuscripts in a number of word-processing formats and produced a coherent text, often at times working above and beyond the call of duty. Joan Hodge, Nikki Lamb and Kay Donaldson all played important roles in the Institute in getting the manuscript to the publisher. Terry Bacon in the Department of Geography at Exeter prepared the final versions of the majority of the diagrams in the book.

David R. Phillips
Yola Verhasselt

Part I

HEALTH AND DEVELOPMENT: PERSPECTIVES AND ISSUES

1

INTRODUCTION: HEALTH AND DEVELOPMENT

David R. Phillips and Yola Verhasselt

The health of populations and individuals is inextricably bound up with development. Development, in the most general meaning of the word, entails change and often important alterations to people's living environments. However, many studies and projects related to development have ignored or skirted around the health dimension. This has been highlighted in recent years by a number of authors and has been brought into sharp focus by the World Health Organization's publication in 1992 of a report of the WHO Commission on Health and Environment, entitled *Our Planet, Our Health* (WHO 1992a). This report takes a systematic view of global change and impacts on food supply, water, energy industry and settlements, and proposes certain strategies and actions. This current book provides very much a complementary perspective, discussing important underlying issues of health change with development and providing specific country and regional studies of health and development.

HEALTH AND DEVELOPMENT

The two terms, beguilingly simple, defy concise and consistent definition. Health is much more than the absence of diagnosed physical disease, although the 'medical model' of medical school training has, often perforce, tended to stress this aspect of health. The constitution of the World Health Organization sees health as 'a state of complete physical, mental and social well-being and not merely the absence of disease or infirmity'. It implies complex interactions between humans and their environments, more particularly between social and economic factors, physical environment and biological environment (Figure 1.1). The disease ecology school of medical geography, developed strongly under May's influence, has recognized this complex, changing and *delicate* interrelationship for many years (Learmonth 1988).

Development, like health, is equally an elusive concept. Many economic definitions see development as being a phenomenon measurable by increasing per capita income or gross national product (GNP). While development

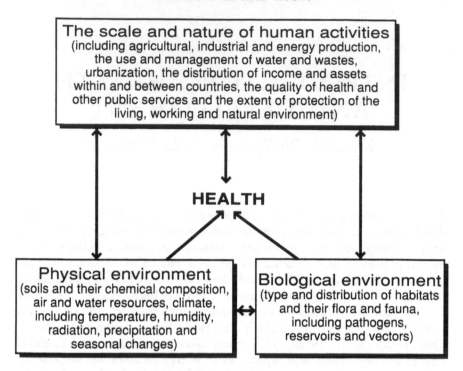

Figure 1.1 Interaction between human activities and the physical and biological
environment
Source: WHO (1992a)

does, undoubtedly, have a quantifiable fiscal element, it has many more
subtle elements, to do with distribution of resources, access to opportunities
(services, jobs, housing, education) and political/human rights. Seers (1969)
sees development as of necessity increasing aspects of human dignity –
which means access to basic needs: food, shelter, a job – and also political
participation. Development is therefore generally understood as the process
of improving the quality of all aspects of human life (WHO 1992a).

A complex interrelationship exists between health and development; it is
certainly not a one-way relationship, and there are surely reciprocal and
synergistic elements to it. It has long been acknowledged that the health
status of the population in any particular place or country influences
development. It can be a limiting factor, as generally poor individual health
can lower work capacity and productivity; in aggregate in a population, this
can severely restrict the growth of economies. On the other hand, economic
development can make it possible to finance good environmental health,
sanitation and public health campaigns – education, immunization, screening and health promotion – and to provide broader-based social care for
needy groups. General social development, particularly education and

literacy, has almost invariably been associated with improved health status via improved nutrition, hygiene and reproductive health. Socio-economic development, particularly if equitably spread through the population – although this is rarely the case – also enables housing and related services to improve. The classical cycle of poverty can be broken by development.

However, it is notoriously difficult to provide generalizations about the relationships between economic development and a population's health status. We cite below examples in which correlations between GNP and life expectancy are not straightforward. There are many examples to show how economic development has contributed to improving quality of life and health status, via indicators such as increased life expectancy, falling infant, child and maternal mortality and enhanced access to services. By contrast, there are examples in which economic development, infrastructure expansion and agricultural intensification do not always coincide with improved human well-being. There is, in fact, growing realization that macroeconomic changes may not always filter down to benefit all of the population, and many perhaps soundly based policies in economic terms – notably structural adjustment policies – can have devastating human effects in increasing poverty and maldistribution of resources. In the face of these policies, the health sector alone cannot overcome concomitant health and welfare problems.

Recently, the WHO has recognized three major groups of problems which have contributed to a growing 'health crisis' in many countries, or which have hindered a clear improvement of health with development (Weil et al. 1990). It behoves the readers of this book to bear in mind these three major problems: first, the very magnitude and diversity of health hazards associated with development; secondly, the cost of diseases caused by industrialization and urbanization and also by medicine itself, the iatrogenic diseases; and, thirdly, the need for, or imposition of, macroeconomic adjustment which has resulted in major cuts in the health and social budgets of many developing – and some developed – countries.

THE IMPACT OF DEVELOPMENT POLICIES ON HEALTH

It is important to recognize that many development policies designed to improve living standards and economic conditions of communities can have unexpected or unintended effects on health. These can be what Phillips (1990) calls the 'health by-products' of industrialization, which include risks from new dangers at work, exposure to toxic substances, effluent, radiation, traffic pollution and general industrial noise and waste in the environment. They can include side-effects of agricultural expansion projects which, at the macro-level, can help nations to feed themselves but which, locally, can bring workers and residents into contact with agricultural chemicals, fertilizers, pesticides and diseases such as malaria or schistosomiasis, which can be associated with unmonitored extension of irrigation. These by-products of development are often reflected in changing epidemiological profiles (the types of diseases from which people suffer). As we discuss below, many poor countries now face a double health burden, from continued prevalence of malnutrition, respiratory infections and diarrhoeal diseases, but also from vastly growing health problems usually associated with industrialization and urbanization: occupational hazards, accidents, cardio-vascular disease, cancers, substance abuse and increasing misuse of pharmaceuticals and quasi-modern cures.

It is therefore important that development projects, whether locally sponsored or supported by international donor agencies, include not only needs assessments but also assessments of environmental impact and health consequences. It is now recognized that, in some sectors, notably in the field of water resource development, engineers and planners have increasingly become aware of potential environmental and health consequences of schemes. A move towards genuine intersectoral co-operation, so often platitudinously stressed, has been possible. However, in many aspects of development policies, the health consequences have been ignored or under-rated. This danger may yet increase in the 1990s, when less co-ordinated, individual and entrepreneurial schemes may be replacing larger-scale bilateral official projects which, in theory, have the scope for more considered assessment of impacts. Weil *et al.* (1990) suggest that the impacts on health of development policies need to be considered in at least five areas outside the health sector, all of which are closely associated with economic growth:

1 Macroeconomic policies, especially adjustment policies, public expenditure, trade policies and household responses.
2 Agricultural policies: irrigation systems; pesticide use; land policy and resettlement; agricultural research and nutrition.
3 Industrial policies: industrial development policies (and location of industry); occupational health policies; control of water and air pollution; management of hazardous wastes.
4 Energy policies: indoor air pollution, domestic fuel use; fuel scarcity and fuel gathering; fuel pricing and energy subsidies; control of pollution from industrial energy sources; health risks presented by hydroelectric power programmes.
5 Housing policies: housing conditions and health; low-income housing policies, slum and squatter clearance; public provision of housing and infrastructure; sites-and-services programmes; upgrading programmes.

All of these five sectors clearly have a potential impact on the health of residents and workers. They are all closely interrelated in many instances, and readers of this book will see evidence in a number of chapters and case-studies of their effects. Many have very direct impacts on the local environment and often have environmental change potential. It is to this important topic that we now turn.

CHANGING ENVIRONMENTAL CONDITIONS AND HEALTH

Environmental change can be of many types and scales. There can be direct change in the local living environment – new housing, factories, reservoirs – or change may be at a much more widespread level, including the impact on health of global environmental change. In addition, what might be called the socio-cultural environment – people's social behaviour, nutritional preferences and consumption habits – is also very important. The understanding of the dynamics of environmental changes of many sorts is now quite good at a theoretical level, but the translation of theoretical knowledge into practical health improvements is not always easy. For example, at a global level, it is known how to reduce or at least contain global warming; but how can this, practically, be achieved? At the personal lifestyle level, in developed countries, there can hardly be an adult or teenager who does not know of the positive and strong correlation between, say, tobacco smoking and lung cancer; yet why do so many people continue to smoke? Individual lifestyle decisions build up into macro-consumer demand, with health consequences to the individual, and health by-product impacts on all. In addition, a mass of research on genetic predisposition to certain diseases is now becoming available, but the precise role of environmental factors as triggers or stimuli is not fully understood, although many associations have been suggested (Foster 1992).

Bentham, in Chapter 2 of this book, reviews the evidence for global environmental change and its potential for health impacts, some of which are already evident. The well-known depletion of the ozone layer and increased exposure to ultraviolet radiation are discussed, and a steady increase in skin cancers, particularly the highly dangerous form, melanoma, is explained. This may be particularly significant in the Southern hemisphere, and Australian health promotion efforts, for example, are now strongly geared to melanoma risk avoidance. Bentham also points to the obvious effects of global environmental change on health, many of which are only now becoming appreciated. A particular health hazard might be an extension north and south of the warm climatic zones in which what are currently regarded as mainly tropical diseases may in future flourish. This applies particularly to vector-borne diseases such as malaria, dengue, yellow fever and trypanosomiasis. Indirect impacts of warmer climates include the potential for increased incidence of food poisoning. Bentham notes that, while there are numerous scare stories associated with macro-scale environmental change, there is no room for complacency. It may be that the exact health sequelae of environmental change may to an extent be speculative, but scientific evidence is growing, and it is certain that increasing serious attention needs to be devoted to this theme. There will surely be both health and economic consequences of environmental change which must be anticipated by us all.

Malaria provides a specific example of a disease that is eradicable but remains one of the most prevalent infectious conditions worldwide, although it principally still affects poorer tropical countries. According to WHO estimates, nearly half of the world population is at risk in more than 100 countries (with an estimated 110 million cases and 270 million people carrying the malaria parasites). It is a debilitating disease and, as such, an obstacle to economic development. It remains a major cause of death (with 1–1.5 million deaths annually), particularly among young children.

Malaria has been eradicated in some parts of the world, for example in Southern Europe. In other regions, however, it remains a major threat and has also seen a resurgence. There are various reasons, mostly concurrent. Among the main causes we can cite loss or reduction of vector control due to the success of previous eradication campaigns and/or to financial restrictions related to structural adjustment; growing resistance of the mosquito vector to insecticides and of the parasite to the cheap and common anti-malaria drugs; the expansion of irrigated land and the development of agricultural schemes increasing population mobility and rapid urban expansion. In the future, global climate change is also likely to increase the area of geographic distribution suitable for the anopheles mosquito. South Asia generally and Sri Lanka in particular witnessed major malarial eradication efforts with notable success after the Second World War. Consequently, in the Dry Zone, where malaria is endemic, agricultural development took place, attracting migrants from malaria-free zones. A major malaria resurgence followed. The role of migration was important, but also financial constraints with regard to vector control and difficulties of internal political disorder are elements in the explanation of the resurgence. In many other countries, malarial resurgence during the late 1980s has been related to a weakening of control programmes. Elsewhere, for example in Brazil and in Rwanda, new land has been brought into cultivation because of population pressure. New in-migrant agricultural workers have been exposed to settings with a high malarial risk. The incidence of the disease has often threatened the success and continuation of such agricultural extension policies.

URBANIZATION = MODERNIZATION = BETTER HEALTH?

The question whether urbanization equates with modernization and better health is an important question in health and development, since many aspects of development theory have held the explicit or implicit assumption that development (in its very many definitions) and modernization will be

8

associated with urbanization. The modernization model, in particular, expected that virtually all economic growth would follow a Western industrial pattern (Hewitt 1992). While there have been valiant experiments to try alternative paths for development on a small scale, and especially for the rural areas in which most Third World residents live, many countries have settled on variants of industrialization strategies as their way forward (Todaro and Stilkind 1981; Kitching 1982; Simmonds 1988; Hewitt et al. 1992). These have almost inevitably entailed greater urbanization. The concentration of people which this entails and many of the industries themselves, by the nature of their activities, their environmental impacts and their effects on society and family life, have added to health hazards compared to more traditional rural and agricultural settlements and activities (Smith 1992).

It is important to differentiate modernization and industrialization. Modernization has historically been perceived as the process of change towards social, economic and political systems that have developed in Western Europe and North America during the seventeenth to nineteenth centuries, spreading elsewhere during this century. Modernization implies a complete transformation of many aspects of life, but it is often associated with industrialization. Broad views of modernization see it as involving interrelated technical, economic and ecological processes: changes from simple traditional techniques towards the application of scientific knowledge; evolution from subsistence farming towards commercial production and specialized agriculture; transition in industry from the use of human and animal power towards industrialization proper and, in ecological arrangements, the movement from farm and village towards urban centres.

In terms of the health impacts of development along urban lines, it has generally been suggested that urban inhabitants will have better health than their rural counterparts because of better access to health services and to cash incomes with which to buy food and services; and, in the past in particular, urban education and health information levels might have been higher. Against this simple proposition is the increasing evidence that health status is rarely directly correlated with the provision of services, and the truth is that many cities can barely provide the basic services for their citizens. Indeed, for the urban poor, many cities do not even provide the most basic level of health services, and other environmental services that would contribute to public health, such as sewerage, potable water and waste disposal, are absent or deficient, as are effective transport and affordable shelter. It seems that often in Third World cities poorer residents are barely any better off than they would be in rural areas, and there is growing evidence that nutrition and family support as well as environment are better in rural areas in many countries. The very scale of both urban and rural poverty in many developing countries is a major constraint on achieving better living conditions and possible improvements in health.

9

Estimates suggest that, at present, an average of 50 per cent of the world's urban population live at the level of extreme poverty and that this rises to 70 per cent in some cities (Stephens and Harpham 1991). Living conditions were rated as only 'fair' or 'poor' in 54 out of the 98 cities surveyed by the Population Crisis Committee in 1989–90. The healthy cities projects have been one international attempt to provide cross-fertilization of good practices and ways ahead in urban development in the future, although practical outputs may be restricted (Ashton 1992).

When the health of urban residents is called into direct question, it appears that urbanization does not automatically equate with better health but it may equate with different health and health problems. It is very difficult to be precise, but when some of the most urbanized societies are considered, particularly some of the small city-states in rapidly modernizing areas such as South-east Asia, epidemiological change of considerable proportions has been noted over the past three decades or so. The major factor that does suggest that urbanization has been accompanied by better health is that there is now much longer expectation of life: indeed, new 'outer limits' to life expectancy are being tested in some modern cities, particularly in East and South-east Asia (Leete 1985; Phillips 1992). The more highly urbanized countries do seem to be those which, on average, have the lowest rates of under-5s mortality and longer expectation of life (Figure 1.2). But a major qualification is that increasing life might paradoxically be accompanied by worsening health, as the main killer diseases are controlled or defeated, but the ailments causing chronic illness have proved much more intransigent.

Urban populations are often at the interface of industrialization and underdevelopment, particularly in Third World countries. The urban poor are now increasingly recognized as a vulnerable group, and this can be in developed or developing countries. The health plight of poor urban residents in cities in India and Peru, for example, may be readily evident. The urban poor in New York and London are increasingly at risk from the chronic dangers of the urban environment and from poor nutrition; if homeless, they may be vulnerable to the weather, to violence, drug abuse and infectious diseases, notably hepatitis, tuberculosis and HIV/AIDS. Thus, as Stephens and Harpham (1992) point out, the urban poor suffer the worst of both worlds in mortality and morbidity profiles. They face the problems of underdevelopment – infectious diseases and excessive infant mortality – and the problems of modern urban industrial populations: illness and death from non-communicable degenerative diseases, neoplasms, cardio-vascular diseases, accidents and neuropsychiatric conditions. Indeed, the last named, mental ill-health, is increasingly recognized as a major source of disability (if not death) in many countries (Harpham 1992; World Bank 1993). This not only afflicts the well-to-do through stress-related and other disorders, but life on the margins of society is also frequently associated with psychiatric morbidity.

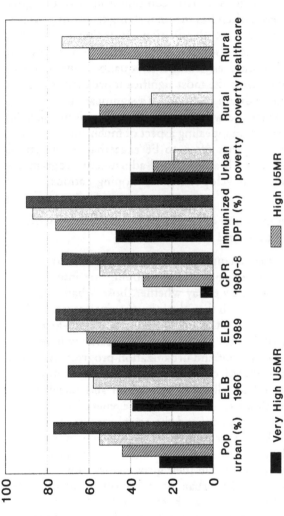

Key

U5MR = Under-5 mortality rate; annual number of deaths of children under five years of age per 1,000 live births.
ELB = Expectation of life at birth.
CPR = Contraceptive prevalence rate; percentage of married women aged 15–49 currently using contraception.
Immunized DPT = Percentage of one-year old children fully immunized with diphtheria, pertussis (whooping cough) and tetanus, 1989.

Urban poverty = Percentage of urban population below absolute poverty level.
Rural poverty = Percentage of rural population below absolute poverty level.
Rural healthcare = Percentage of rural population with access to health services.

Figure 1.2 Urbanization and health indices: countries by under-5 mortality rate
Sources: UNICEF (1991); UNDP (1992)

Three main groups of diseases are defined by Stephens and Harpham (1992): diseases of poverty; environmental diseases of industrialization; and psychosocial diseases of industrialization. By implication, these all have direct associations with, and are negative by-products of, development and industrialization. There are admittedly considerable difficulties in attempting to organize what are likely to be highly interrelated and complex health problems into discrete categories. Many conditions involve a lengthy exposure or risk period, but each often has the underlying factor of poverty, which is itself a composite index of deprivation. It includes not only deprivation relating to economic resources but usually, too, access to education, social support, housing, environmental quality, nutrition and many other factors. Poverty remains the most significant predictor of urban morbidity and mortality, and no doubt it also plays a significant part in rural ill-health. Increasing divergence between rich and poor within specific countries and cities is a major and unappealing aspect of much contemporary development, and one which appears often to be exacerbated by official action (or inaction) and policies such as structural adjustment programmes or demands of debt servicing (in developed and developing nations).

Epidemiological transition: some examples of health change with development and modernization

In the context of epidemiological change, it is no longer suitable to classify cities or countries simply into developed or developing, or into North–South or East–West. Neither is it easy to say whether those urban areas that have grown rapidly have 'better' health than the less urbanized areas. Rather, it is perhaps more accurate to say that there will be various sub-groups in most populations in terms of health. In addition, ill-health will tend to assume different forms as development and urbanization progress. As noted above, many so-called Third World or developing countries today have epidemiological profiles that reflect all types of medical and social needs: infectious and parasitic conditions; chronic and degenerative diseases; psychological and psychiatric morbidity; and the social care needs of very young and very old people. A number of authors have begun to point out the necessity of looking closely at the epidemiological conditions *within* specific countries and, particularly, within their growing urban populations and specifically their impacts on the urban poor (Harpham *et al.* 1988; Hardoy and Satterthwaite 1989; Tabibzadeh *et al.* 1989; Hardoy *et al.* 1990). It is becoming increasingly evident not only that there are differences over time in the nature of diseases in countries but that their internal distribution can be quite variable. This can reflect differences among economic, social or occupational groups, between regions or districts or between town and country (Harpham *et al.* 1988; Frenk *et al.* 1989; Phillips 1990, 1991).

The general concept of epidemiologic(al) transition

The basic principles of epidemiological transition and the relationship of this proposed transition with the better-known demographic transition have been outlined by authors such as Omran (1971, 1977) and Caldwell (1982). The idea of epidemiological transition is quite straightforward, and the 'theory', as Omran (1977: 4) calls it, 'focuses on the shifting web of health and disease patterns in population groups and their links with several demographic, social, economic, ecologic and biologic changes'. The 'theory' addresses the nature of the relative balance between various causes of mortality in particular (and morbidity implicitly) and the ways in which changes occur while societies modernize. Today, the term *health transition* is preferred by some, and was used in an exploratory Rockefeller Foundation programme. Health transition implies the cultural, social and behavioural determinants of health and implies a concern with health and survival rather than death; it also suggests continuing, socially influenced change (Caldwell *et al.* 1990; Hill and Cleland 1991; Chen *et al.* 1992; Caldwell 1993). The concept of 'risk transition', which emphasizes the evolution of environmental and occupational health hazards, may also be helpful (Smith 1990).

The idea that behaviour and lifestyle can play a major role in determining health and risk of death is certainly not new. The precise balance of the contributions of lifestyles, public health measures, medical care and general improvements with development to health, however, is quite fiercely debated. This book shows in many chapters how lifestyles and behaviour can be critical influences on health, particularly in certain environmental contexts. Bentham (Chapter 2) outlines the changing environmental risks of development; Ford (Chapter 5) discusses the crucial role of behavioural factors in HIV transmission; Price (Chapter 9) discusses the role of mothers in children's health.

Epidemiological transition as originally envisaged involves a one-way movement in the main stages from a preponderance in the Old World of epidemics of infections and famine, through an era of receding pandemics, to a pre-eminence of degenerative or sometimes 'man-made' (human-made) diseases. Some authors are implicitly and explicitly discussing a fourth stage of epidemiological transition, in which length of life expectancy increases (as major killer diseases are being better treated or detected) but in which health may deteriorate, as the causes of chronic but non-fatal morbidity are yet to be defeated (Verbrugge 1984; Riley 1989; Riley and Alter 1989; Phillips 1991). Epidemics or the increasing incidence of mental disorders seem also to be characteristic of this fourth stage. Olshansky and Ault (1986) have called this the 'age of delayed degenerative diseases' and see it as a stage that will propel life expectancy into and perhaps beyond the eighth decade. This is already evident in a number of developed countries and in some middle- and upper-income newly industrializing countries, particularly in

South-east and East Asia (Leete 1985; Phillips 1992). It is clear in the context of this book that such population ageing has crucial implications for employment prospects and for service demands, as Warnes discusses in Chapter 10 and Joseph and Martin Matthews in Chapter 11.

The 'speed' of the transition and the factors influencing it, however, appear to have varied from one group of countries to another. In the West, transition began before many modern medical discoveries such as antibiotics and was clearly associated with improved standards of living and public health (McKeown 1965, 1988). The transition in the West is now probably almost, if not quite, complete. It took 100 to 200 years and is the classical or Western variant of the theory. In its first stage, the 'age of pestilence and famine' gradually merged into a period of 'receding pandemics', to be followed in the mid-twentieth century by an 'age of degenerative and man-made diseases'. This transition tended to be smooth and gradual. It is generally argued that, as it was associated with the industrial and social revolution in the West, it clearly cannot be exactly repeated in other parts of the world.

The accelerated transition is the second variant of the model and may be seen in countries such as Japan and the former Soviet Union and probably also in certain others in Eastern Europe and South-east Asia. Mortality and fertility declined rapidly, and a rapid change took place to a modern epidemiological profile. By contrast, some countries, principally in the poorer Third World, appear to be in a 'delayed' model. The transition started late, but was effected to an extent by Western technology. In so far as fertility has not always declined rapidly and living conditions have not always improved substantially for all people in these countries, their transitions have not been complete. Many aspects of degenerative and man-made diseases have become evident, but many infectious ailments remain. These countries (or, at least, certain of their citizens) in effect may suffer from the worst elements of both major groups of ailments. Omran (1971) singled out Sri Lanka as an example of the delayed transition, but today certain countries in Africa are likely to be cited, as well as some in South Asia.

It is perhaps the middle-income countries today that are experiencing the most rapid changes in health in some areas and for some sectors of their populations. This implies, to Phillips (1988), Frenk et al. (1989) and others, a modification of the theory, or at least of the third stage, in which there is not a simple sequence of the eras but a period in which two or more may overlap. In addition, it appears that changes in patterns of morbidity and mortality may be reversible, and that a type of counter-transition may occur. It would seem that this is the case in countries which still suffer from outbreaks of infectious diseases and famines (which are often human-influenced). Peru in the early 1990s provides an example of a country with relatively high morbidity and mortality from cholera, spreading to other parts of Latin America. Many parts of Africa are experiencing food shortages and morbidity and mortality patterns reminiscent of the early eras of epidemiological transition. In Chapter 13, Kloos points out that this is clearly evident in the poorer Third World, while Akhtar and Izhar (Chapter 14) and Curto de Casas (Chapter 15) show evidence of epidemiological change in India and Latin America.

It is also likely that a see-saw effect of health improvements occurs within specific countries or in parts of countries. Sometimes only certain sectors of the population (the very young; mothers; the poor; or infirm people) may be affected. This coexistence has been called a protracted epidemiological transition (Frenk *et al.* 1989). In some instances, there is growing evidence of epidemiological polarization. Richer sectors of populations may develop more or less 'modern' health disease profiles; some poorer sections may experience infectious and nutritional disorders on top of, or instead of, more degenerative and chronic ailments. The inhabitants of many cities in developing countries, in particular industrializing ones, will often be exposed to 'traditional' infectious and environmental health risks as well as to chronic ailments and the spin-offs of industrialization. The social factor as a determinant of health status is in many ways gaining enormous influence in these cities in a way that it has ceased to do, or that is much muted, in Western cities. Who you are and which part of the city you live in become major factors influencing the health and life chances of you and your family. This obviously has very important implications for public health policy and for the planning of health and social care services.

It might be possible to use the concept of epidemiological transition to provide a basis for planning to meet the type and scale of future health care needs that might reasonably be expected within various countries. Many authors have suggested that this is a potential approach (McGlashan 1982; Hellen 1983; Phillips 1988, 1990; Lopez 1989; Picheral 1989). However, data limitations outlined below may render it impractical. It may also be unwise to use a model which assumes a linear unfolding of successive stages in what is certainly a very complex relationship (Jones and Moon 1992). Whatever one feels, however, it is essential to look to the future. Many forms of health care provision and health education campaigns have long 'lead times'. Primary health centres may be provided relatively quickly, but new hospitals require many years of planning and building, and in both cases the training of staff for the diseases and conditions they will meet in the future is more important than an over-concentration on those conditions they may have met in the past.

Epidemiological change in middle-income countries

Many countries in the middle-income group appear to have gone relatively rapidly through some elements of epidemiological transition. This is especially

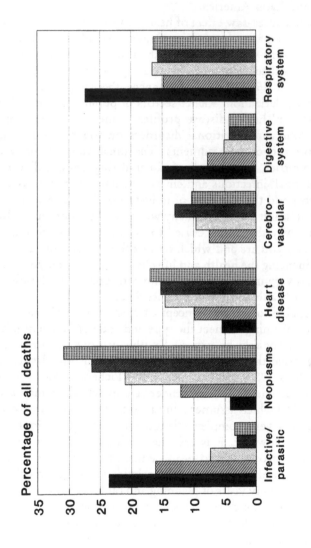

Figure 1.3 Hong Kong: epidemiological change, 1951–91 (% deaths)
Sources: Phillips (1988); Hong Kong government figures (1991)

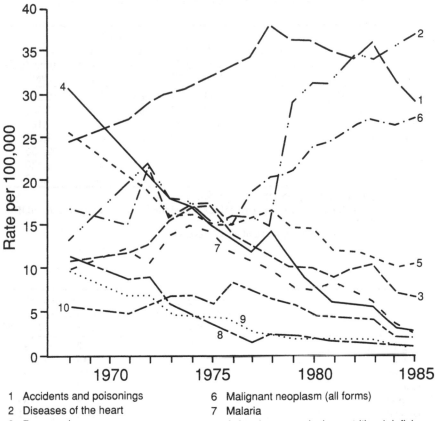

Figure 1.4 Health transition in Thailand: crude death rates per 100,000 of population
for major causes of death, 1968–85
Source: Institute for Population and Social Research (1988)

true in parts of South-east Asia and Latin America, although Hansluwka
(1986) considered that, in the majority of countries, by the mid-1980s the
synergistic interaction of malnutrition with infectious and parasitic diseases
was still paramount, with malaria, tuberculosis, leprosy and acute diarrhoeal
diseases maintaining considerable importance (the last named particularly
for infants and children). Three or four countries in South-east Asia do none
the less stand out as examples of this group which have experienced fairly
rapid, and in some cases spectacular, transitions. These are Hong Kong,
Singapore, Malaysia and Thailand. In these countries, modern epidemiological profiles have been emerging, over the period after the mid-1960s in

Table 1.1 Health indices in various middle-income countries

(A)

	Population urbanized (%) 1990	USMR (per 1,000) 1960	USMR (per 1,000) 1990	Life expectancy at birth (years) 1960	Life expectancy at birth (years) 1990	GNP per capita (US$) 1989
Hong Kong (HHD)*	94	64	7	66	77	10,350
Jamaica (MHD)	52	89	20	62	73	1,260
Malaysia (MHD)	43	105	29	54	70	2,160
Mexico (HHD)	73	140	49	57	70	2,010
Singapore (HHD)*	100	49	9	65	74	10,450
Thailand (MHD)	23	149	34	52	66	1,220
Turkey (MHD)	61	258	80	50	65	1,370

Sources: World Bank (1991); UNDP (1992); UNICEF (1991)
Note: HHD = high human development; MHD = medium human development (UNDP categories, 1992)
* now high income economies but classified as 'developing'

(B)

	Infant mortality rate (per 1,000 live births) 1990	Percentage of 1-year-olds immunized 1981	Percentage of 1-year-olds immunized 1989–90
Hong Kong	7	92	81
Jamaica	16	38	86
Malaysia	22	69	93
Mexico	40	50	78
Singapore	8	79	89
Thailand	26	48	91
Turkey	69	57	72
Average for *least* developed countries	115	17	57
For *all* developing countries	74	24	79

Source: UNDP (1992)

(C)

	Contraceptive prevalence (%) 1985–7	Maternal mortality 1980–7	Adults who smoke (%) 1985	Population per Doctor 1984	Population per Nurse 1984
Hong Kong	72	5	19	1,070	240
Jamaica	–	110	–	2,040	490
Malaysia	51	59	29	1,930	1,010
Mexico	53	82	32	1,240	880
Singapore	–	5	29	1,310	–
Thailand	66	–	36	6,290	710
Turkey	77	210	50	1,380	1,030

Source: UNDP (1991)

particular. Elsewhere, Jamaica, Turkey and Mexico typify middle-income countries. Thailand and, to an extent, Malaysia, Mexico and Jamaica retain important elements of mortality from infectious conditions and accidents. Figure 1.3 shows the distinct trend towards a modern epidemiological profile for Hong Kong during the course of this century and particularly since the 1960s. Figure 1.4 shows the increasing relative importance of heart disease and cancers in particular, and the decrease of most infectious ailments, as causes of mortality in Thailand over the past few decades. Mexico has seen very marked falls in mortality from malaria, whooping cough and dysentery in the period since 1950, and also steep rises in mortality from ischaemic heart disease, diabetes and motor vehicle accidents. Yet, like many other countries in this group, Mexico is in a type of protracted transition, of the kind noted earlier.

While in many of the countries there may be a steady epidemiological change on average, some, such as Mexico, Thailand and Malaysia, continue to have groups or pockets of population with high infant mortality and relatively high incidence of infectious disease mortality or morbidity. The under-5 mortality rates in Table 1.1 show considerable variety both in their current (1990) levels, ranging from 7 per 1,000 births in Hong Kong to 80 in Turkey, and in the rate of decline over the past thirty years or so. Infant mortality today also shows a considerable range, as do proportions of 1-year-old children immunized. There has been a general increase in the life expectancy at birth, but this has reached higher levels in Hong Kong and Singapore (mid- to upper seventies), than in Thailand and Turkey (mid-sixties). There is not a simple, one-way correlation between life expectancy and its rate of increase and (say) GNP per capita. The WHO (1992a) and Caldwell (1993) cite examples such as Sri Lanka, China and Vietnam, with very low per capita incomes yet life expectancies of around 70 years, which contrast with upper-middle to high-income countries such as Saudi Arabia, Libya and Oman, where life expectancy may be five to ten years lower. Success in extending life expectancy in the first group was seen as a result mainly of the exercise of political and social will, especially access to good community health programmes and a high commitment to female schooling. While there is a general relationship between the purchasing-power-based measure of per capita GDP and life expectancy, this is only one measure among numerous socio-cultural factors and others that influence health. There are many exceptions, among the most spectacular perhaps being Cuba, Jamaica, Sri Lanka and China, in which life expectancy is far ahead of what might be predicted based solely on economic criteria, which cautions against simple cause-and-effect explanations (Figure 1.5).

Tables 1.1(B) and 1.1(C) also indicate that there is not a simple correspondence between crude indicators of health care availability (highly selectively chosen here) and either life expectancy, maternal mortality, infant and under-5 mortality or other indicators of health status. There are huge

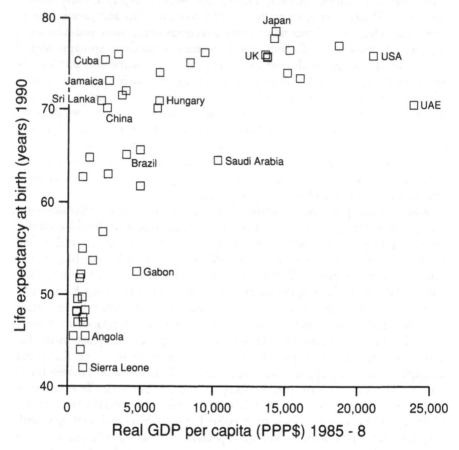

Figure 1.5 Countries' life expectancy plotted against real per capita GDP for latest available year

variations between countries which have similar ratios of health personnel to population, which means that many other factors, too numerous to cover here, must be involved in the study of epidemiological change and health care needs. To basic indicators of population structure, distribution, morbidity and mortality there need to be added sensitive socio-cultural indicators and distributional indices.

There is a manifest need for detailed within-country study of epidemiological trends, as well as comparative cross-national research. The importance of within-country research is that it may be able to highlight, for example, cultural differences, socio-economic and ethnic variations, local needs and localized maldistributions of resources. Aggregate research cannot do more than hint at such variations.

Problems of data

The availability of reliable, good-quality data over a sufficiently long time period and with a comprehensive geographical spread within specific countries is perhaps *the* major factor permitting or limiting research in many countries, at whatever stage of development. In many countries, neither good nor comprehensive mortality data exist, particularly over a time span sufficient to detect substantial change. Many countries also rely mainly on crude rather than accurate age-adjusted mortality data, as epidemiological data by age groups are not reliably known.

Quality of data can be affected by many factors. For example, geographical and temporal variation in apparent mortality levels may be caused by any one of a combination of factors, some of which are more or less recording aberrations. These occur as statistical artefacts (diagnostic habits of certifiers and of categorizations on death certificates) or through disease incidence (varying with climatic, demographic, socio-economic and nutritional variables) and case fatality which relate to an extent to individual help-seeking behaviour, quality of care and health status. Certification and recording of data are also often suspect, particularly the further back in time one goes. Therefore, any research must recognize the limitations imposed by data quality, and be circumspect in the interpretation of findings. When international mortality data are being considered, the bases on which these have been collected and analysed become crucial, as rarely are reliable and truly comparable mortality data available for developing countries.

Reliable epidemiological data, with fairly comprehensive coverage, will be needed on morbidity as well as mortality if firm statements are to be made about relationships between health and development factors. Data also need to be collected and analysed on a basis that will enable gender, ethnic, socio-economic and regional differentials (among others) in health and health needs to be identified. It is increasingly likely that changes in morbidity, as much as, or more than, changes in mortality will be of major significance to future demands for health and social care. Morbidity data can often reflect the major reasons for contact with, or need for, health services, as the actual cause of death may be different from the main causes of illness in an individual's lifetime.

There are many reasons why morbidity data become of increasing importance as epidemiological transition progresses, but a principal one is that, as chronic or degenerative diseases increase in importance, they may be associated with considerable amounts of ill-health, restricted activities and need for health and social care. Underlying diseases or degenerative diseases may be more important in terms of human suffering and needs for health services than the actual diseases that ultimately cause death. Verbrugge (1984) has noted the phenomenon of 'lengthening life but worsening health', as more causes of mortality are being dealt with while

21

non-fatal but disabling conditions receive relatively less attention and are much more intractable.

As far as morbidity data are concerned, the picture is gloomy. Very few reliable data on morbidity are available in many countries even today. Most available information is based on case studies which are generally not useful for plotting change over time but which provide a quasi-static 'snapshot' of what is probably a very dynamic phenomenon. Even the Demographic and Health Surveys (DHSs) of the 1980s and 1990s present rather restricted sample data on morbidity, in part due to limited knowledge on the part of the respondents. The DHSs also have a concentration on mothers' and children's health. This general lack of data on *morbidity* change is perhaps one of the greatest weaknesses in attempting to research epidemiological change and in the understanding of the relationships between health and development.

KEY CURRENT ISSUES IN HEALTH AND DEVELOPMENT

A number of critical issues in health and development have emerged in the late 1980s and early 1990s. Many concern vulnerable groups, the health of the urban poor or the health of women in so far as their health needs have been subordinated to those of children; victims of political violence and health of refugees; and the impacts – intended or unintentional – of development policies, donor aid, structural adjustment and cost recovery programmes. Other concerns involve excessive curative-technology orienta-tion of much contemporary health care in the majority of countries. Most of these concerns are elaborated in the following chapters, but it is useful to provide an introductory statement on key issues at this stage.

Health impacts of adjustment policies

A major characteristic feature of almost all economies over the past decades or more has been retrenchment and cut-backs in public-sector involvement in services. These reductions impinge on health and have occurred at times of recession when the private sector, even should it have wished to do so, has been unable to step into the breach. Neo-liberal development policies have provided the dominant view during the 1980s, and these policies have emphasized the individual and the free play of market forces. Economic reform (mainly involving a reduction in the role of the state) and structural adjustment programmes, which have aimed to 'readjust' many developing economies, have almost inevitably in the immediate term involved lowering standards of living and reducing state expenditure on sectors such as health, welfare and education. In Chapter 3, Asthana discusses in more detail the actual and potential effects of these policies on health, so it is only necessary to introduce this important issue here.

Structural adjustment policies have often sought to force governments to restructure their own taxation and social provision. The poor and groups such as children and disabled or elderly people tend to be particularly vulnerable to cuts in services (Messkoub 1992). Their health chances will inevitably be impaired, and recognition of this has led to calls for 'adjustment with a human face', to raise the consumption levels of the poor to the basic needs minimum, at times when constraints on consumption levels are greatest (Weil *et al.* 1990). It is probable that the impacts of economic adjustment policies in health terms will be greater in urban than in rural areas, as urban-sector programmes and provision have tended to be more heavily underpinned by public funds than programmes and provision are in the rural areas, where extended families are more often alleged to be able to take care of their members in need. Adjustment programmes can also reduce the scope for environmental health improvements in urban areas, and earlier public health advances may actually be lost (for example, in sewage programmes or malaria eradication). Increased food prices and costs of living which often result from abolition of subsidies and price controls impact greatly on the urban poor and can worsen their diets and living conditions, while services to help them deteriorate or disappear. Cost recovery schemes, for example in the provision of drugs, may mean that the poorest citizens do not always receive the medicines they need, because they cannot afford them (as opposed to the lack of availability of many essential drugs, which is apparent in many developing cities).

Structural management changes can also make health gains difficult. Central government ministries such as those involved in health or education may fail to work well with local authorities charged with, say, environmental improvements, pollution control or housing. This lack of intersectoral co-operation can be a major stumbling block to practical and sustained gains, particularly in primary health care. Much-needed action at the local level therefore may be weak or non-existent. Decentralization is often paid lip-service, but in practical terms its effects may be limited. If it *is* undertaken, power and resources are often only given up reluctantly by higher-tier authorities. Adequately trained personnel may not exist or may not be tempted to work at lower tiers which often retain a low-status image. Co-ordination may be replaced by competition or jealousies between levels of government, and health can suffer as a result.

This suggests that political will is often lacking to allow or encourage health and other related initiatives to succeed. It may be that there is a reluctance on the part of certain officials to encourage and work coherently towards health-related goals. Private-interest rather than public-interest views of the state can help in the analysis of this phenomenon.

In this discussion, an important constraint is that it is very difficult to demonstrate direct health impacts of many elements of change, and this can deter sustained investment. It appears in general, however, that health and nutrition can be adversely affected by economic adjustment programmes, although quantification of impacts is difficult, and some studies are inconclusive. A lack of baseline studies often precludes efficient evaluation of, say, slum improvement projects and particularly their health impacts. Longitudinal studies of health changes are badly needed in many areas, but these are expensive and difficult to implement, particularly over a time scale sufficient to allow the investigation of local epidemiological changes. Stephens and Harpham (1991) indicate that slum upgrading projects, for example, may have very beneficial effects on the physical condition of many slum *areas*. However, their effects on *individual* slum households are often ambivalent and more difficult to analyse. In addition, it is not known what happens to the health of those that urban improvement schemes miss or even 'force out'.

The Bamako Initiative: strengthening local health services

Some countries and continents have attempted to address spiralling health costs and to revive and extend peripheral services locally, and enhance the availability of essential drugs (Phillips 1990). The most notable attempt has been the Bamako Initiative, starting in 1987, to institute some form of cost recovery from health service users and to establish rolling funds to finance and sustain health services. District-level strengthening of services has been seen as crucial, and it has been felt that people will be willing to pay at least something for good-quality services and medicines. Critics feel that the poor cannot afford even minimum expenditure and that all health care should be free. Advocates of the initiative say that it is part of adjustment with a human face, and that it has had some success in extending primary health care (PHC). Roughly half the population of Africa has no access to health care, and much free care has been focused in urban areas. Closing this gap has been part of the Bamako Initiative, and has had some success in rural areas. This is probably patchy, however, although individual countries have reported successful establishment of revolving funds for drug supply (Jarrett and Ofosu-Amaah 1992). The clear problem remains the extension of services to those who have none at all, and cost recovery schemes will probably have limited impact on this problem.

The scale of health problems

In this respect, the very *scale* of environmental problems may in the end prove to be one of the major constraints on health advances, particularly in the poorest countries. In cities in the developed world it is also evident that various forms of urban decay may affect health, and that psychosocial problems, substance abuse, and the like, arise in pockets in affluent countries. There is evidence of this in many North American and European cities at present.

24

A key constraint in the future will be that, while 'lifestyle' changes to improve health may be within the *theoretical* scope of individuals and families, the socio-economic changes needed to effect environmental improvement are in practice not. Individuals are generally subject to forces beyond their own control. For example, unemployment due to circumstances beyond their influence may lead families to poor housing and nutrition, and to physical and psychological health problems. Inflation, health care cost explosions and other economic difficulties may render the poor – and increasingly the not-so-poor – less able to pursue healthy lifestyles or to live in healthy ways. Thus, poverty itself is a major constraint, and it is highly likely that *reverses* in earlier health gains will be seen, with the resurgence of infectious diseases and continuation of disease related to poverty.

Women's health

The health of women, particularly in rapidly developing areas, has become a major concern. Until recently, maternal and child health (MCH) programmes have tended to focus on children and on reduction of infant mortality, the silent emergency identified by UNICEF (the United Nations Children's Fund) (Abou Zahr and Royston 1991). However, with the increasing recognition of health as a matter of family and holistic concern, and the realization that healthy women are a crucial component in the health of nations, the focus has shifted somewhat, if insufficiently, on to women's health. Lewis and Kieffer, in Chapter 8, discuss this in detail. The health needs of women go beyond mother-and-child needs and include reproductive rights, political participation and social equality. These are needed not only for general health reasons but also to enable women to escape abuse, exploitation and infections such as HIV/AIDS (Wallace and Giri 1990; Richters 1992).

Elderly people

Elderly people form another major group whose needs come to the fore with development. In many developed countries, retired people (usually aged over 60 or 65 years) tend to form from 11 to 15 per cent or more of the total population. A particular feature has been the increase in the proportion of older retired people, aged 75-plus. Their health care and social care needs are often greater than those in young age groups; and, while old age by no means automatically equates with disability, this group often does have certain specialist secondary, primary and community care needs. In most developing countries older people currently form a relatively small proportion of the total, typically between 2 and 6 per cent of the population. However, their absolute numbers are often very large (take, for example, China's elderly population, which at 6 per cent of 1,150 million is a huge number). Therefore, both by virtue of the increasing number of elderly people in many Third World countries, and because of their absolute totals, it is essential to start to plan for their future health care, social care and housing needs (Chen and Jones 1989; Tout 1989; Phillips 1992).

Warnes (Chapter 10) and Joseph and Martin Matthews (Chapter 11) provide developing- and developed-world perspectives on health, health care and economic implications of population ageing. This is highly likely to become a key issue in health and development research in the future, because not only are numbers of elderly people increasing, but, with this, there are also the shifts in the epidemiological profile identified earlier in this chapter.

The effects of war and political violence

Many groups and individuals of all ages, ethnic backgrounds and nationalities have been victims of war or political violence. Direct and indirect injuries, both mental and physical, occur very widely in many parts of the world, a sad commentary on the links between health and development. Research into political violence – broadly defined to include war, militarization, repression and suppression of human rights, including the use of torture – has made it now recognized as a major health threat in numerous countries. Many people in South Asia have suffered in the recent past; many in Africa and Latin America continue to suffer from armed conflicts or from oppressive government regimes. Intransigence is often the worst in quasi-civil wars, such as in the 1990s in the former Yugoslavia; but smouldering violence and political and religious intolerance, as is continuingly evident in Northern Ireland, also do untold damage to physical and psychological health. Zwi and Ugalde (1989) have produced a listing of the effects of political violence, with a particular reference to their impacts on health services, reproduced in Table 1.2. The disruption to care services, alongside the enhanced need for both curative and health-promotive services, as a consequence of war and conflict is a major problem in many countries and regions (Phillips 1990).

In addition, as Table 1.2 indicates, refugees are often fleeing from political violence; and the health care needs, physical and emotional, of displaced people now create a major financial commitment for many governments and agencies. Malawi, for example, has been host to hundreds of thousands of refugees from Mozambique, overwhelming the nation's limited ability to provide health care and shelter, and necessitating massive international agency assistance. Similarly, the mental and physical health of Vietnamese refugees has been the subject of concern, and Thailand has for years perforce been host to refugees from Cambodia. These human-misery side-effects of conflict appear to be an alarmingly increasing aspect of many so-called development struggles. Their health impacts are but one aspect, tied up inextricably with economic, physical and political disruption.

Table 1.2 Health and violence

(A) *Effects of political violence*

Direct effects of violence:
Death
Disability
Injury
Destruction of health services (see Table (B))
Disruption of health programmes (see Table (B))
Psychological stress

Indirect effects of violence:
Economic pressures and disruption
Decreased food production distribution
Militarization and diversion of funds to military needs
Family destruction, increased number of orphans, abandonment of children
Displacement of people
Refugees
Psychological stress
Impact on housing, water supply and sewage disposal
Economic pressure of caring for those disabled by violence

(B) *Impact of political violence on health services*

Disruption of lines of communication and referral
Diminished training and supervision of staff
Physical isolation of services from each other
Increased difficulty of attracting staff to work in peripheral areas
Increased difficulty in gaining access to services, e.g. due to curfews
Disruption of mass campaigns, e.g. immunization
Disruption of case-finding and community care
Disruption of routine monitoring and surveillance measures
Reduced infectious disease control measures, e.g. malaria control
Lack of supplies, including drugs and equipment
Emigration of skilled health personnel
Greater dependence on foreign personnel and aid
Diversion of resources to providing acute care for injuries
Long-term demand on health services for providing rehabilitation services
Death, assaults and repression of health workers
Destruction of health facilities

Source: Zwi and Ugalde (1989: 634–5)

High-tech care and costs

An important implication of modern development for health care is an apparent feeling in many countries that high-technology, hospital-oriented care is required. There is already a well-recognized cost explosion in health care prices, particularly for imported high-tech diagnostic and treatment facilities and for pharmaceutical products and medicines. This is in many instances a false trail of development, and in many developed countries there is an on-going attempt to reinvigorate or to boost primary-level care and/ or community care. The long-standing WHO commitment to primary health care as the principal way of assisting with Third World health needs requires emphasis. Indeed, recent WHO publications illustrate the cost-effectiveness of community services as opposed to curative health care in terms of lives saved (Table 1.3). However, the proportion of health budgets spent on such services is only a small fraction of what is spent on treatment and technology. In this book, Price in Chapter 9, Asthana in Chapter 12, Kloos in Chapter 13 and several other contributors draw attention to the need to support primary and community services and to reorient attitudes of planners and medical practitioners to this direction. This must inevitably entail amendments to medical- and nursing-school curricula in many countries, so as to give enhanced professional and career recognition to the primary and community level of care.

PROSPECT

This book does have many positive messages to communicate about health and development although many images portrayed so far have been depressing: poverty, infections, ageing, overstretched services, and the like.

Table 1.3 Spending for and cost of various health services

Services	Percentage of total expenditure on health[a]	Approximate cost per additional life saved (US$)
Direct services to patients:		
Curative: treatment and care of patients through health facilities and independent providers (including traditonal practitioners); retail sale of medicines	70–85	500–5,000
Preventive: maternal and child health care (e.g. immunization, growth monitoring, family planning, promotion of better breast-feeding and weaning practices); adult care (e.g. hypertension screening, pap smears)	10–20	100–600
Community services:		
Vector control programmes; educational and promotional programmes on health and hygiene; monitoring of disease patterns	5–10	<250

Source: WHO (1992b: 46)
[a]Includes both non-governmental and public spending. Data on private spending are available for only a limited number of countries. Figures here assume 90 per cent of private spending is for curative care and 10 per cent is for preventive care.

In many ways, epidemiological transition should be regarded as a major human achievement, as should population ageing. The challenge, to which readers of this book may wish to respond, is to determine and work towards provision of sympathetic, effective human services. If this involves economic re-orientation or a redistribution of resources among countries and populations, then this will have to be addressed. Unfortunately, the last decade of the twentieth century appears set on a less altruistic path, one which threatens to continue to divide and on which good health and adequate health services will be the preserve of the lucky or the well-to-do.

The World Bank's *World Development Report 1993* published after this book had been written, reinforces many of the points made by the authors. The report's theme 'Investing in health' focuses on the interplay between human health, health policy and economic development. The report has three main messages, principally for developing countries. First, an environment should be fostered which enables households to improve health. To this end, governments should pursue sound macroeconomic policies that emphasize the reduction of poverty. Expanded investment in schooling, particularly for girls, is advocated, whilst the rights and status of women should be enhanced, through political and economic empowerment and legal protection. Second, government health expenditure should be made more effective and reorientated towards helping the poor through low-cost, high effectiveness programmes, basic public health programmes including infectious disease control via immunization and AIDS prevention, and essential clinical services, including family planning and reproductive health. Government health services management must be improved and responsibilities decentralized. Imbalanced expenditures on tertiary care and less cost-effective services should be reduced. Third, competition and diversity, especially involving carefully regulated private sector competition for public services is advocated. Whilst it is possible to be concerned about some of the report's financial emphases and about excessive private sector devolution of some health provision, it does make numerous telling points especially for the poorer countries. The quest for effective use of scarce financial and other resources is crucial. The economic and social consequences of continuing irrational expenditure on health are profound. The report also introduces a measure of the burden of disease, rather than the cruder measure of mortality which rarely indicates accurately the personal, social and economic consequences of prolonged ill-health. The measure, disability-adjusted life years (DALY), combines healthy life years lost because of premature death and as a result of disability. The comparison of these values between countries, regions and, potentially, between groups and areas within countries may permit investigation of the real impacts of epidemiological transition and its implications not only for health services but for social care and family support.

STATISTICAL SOURCES

A range of statistical sources on health and development exist, and these often include narrative about important current issues. Among the more useful and accessible sources are:

Demographic and Health Surveys ⎫ Reports available for a large number of
World Fertility Survey ⎬ countries, mainly in the Third World.
United Nations Development Programme (annual) *Human Development Report*, Oxford: Oxford University Press for UNDP.
United Nations Children's Fund (annual) *The State of the World's Children*, Oxford: Oxford University Press for UNICEF.
United Nations *Demographic Yearbook*, New York: UN.
United Nations (biennial) *World Population Monitoring*, New York: UN.
World Bank (annual) *World Development Report*, New York: Oxford University Press.
World Health Organization (annual) *World Health Statistics Annual*, Geneva: WHO.

REFERENCES

Abou Zahr, C. and Royston, E. (1991) *Maternal Mortality: A Global Factbook*, Geneva: WHO.
Ashton, J. (ed.) (1992) *Healthy Cities*, Milton Keynes: Open University Press.
Caldwell, J.C. (1982) *Theory of Fertility Decline*, London: Academic Press.
—— (1993) 'Health transition: the cultural, social and behavioural determinants of health in the Third World', *Social Science and Medicine* 36(2): 125–35.
Caldwell, J.C., Findley, S., Caldwell, P., Santow, G., Cosford, W., Braid, J. and Broers-Freeman, D. (eds) (1990) *What We Know about Health Transition: The Cultural, Social and Behavioural Determinants of Health*, 2 vols, Canberra: Australian National University Press.
Chen, A.J. and Jones, G. (1989) *Ageing in ASEAN: Its Socio-Economic Consequences*, Singapore: Institute of Southeast Asian Studies.
Chen, L.C., Kleinman, A. and Ware, N. (eds) (1992) *The Health Transition*, Cambridge, MA: Harvard University Press.
Foster, H.D. (1992) *Health, Disease and the Environment*, London: Belhaven Press.
Frenk, J., Bobadilla, J.L., Sepulveda, J. and Cervantes, M.L. (1989) 'Health transition in middle-income countries: new challenges for health care', *Health Policy and Planning* 4(1): 29–39.
Hansluwka, H. (1986) 'Mortality in South and East Asia: an assessment of achievement and failure', in H. Hansluwka *et al.* (eds) *New Developments in the Analysis of Mortality and Causes of Death*, Bangkok: Institute for Population and Social Research, Mahidol University, and WHO, pp. 325–408.
Hardoy, J.E. and Satterthwaite, D. (1989) *Squatter Citizen: Life in the Urban Third World*, London: Earthscan Publications.
Hardoy, J.E., Cairncross, S. and Satterthwaite, D. (1990) *The Poor Die Young*, London: Earthscan Publications.
Harpham, T. (1992) 'Urbanization and mental disorder', in D. Bhugra and J. Leff (eds) *Principles of Social Psychiatry*, Oxford: Blackwell, pp. 99–121.
Harpham, T., Lusty, T. and Vaughan, P. (eds) (1988) *In the Shadow of the City: Community Health and the Urban Poor*, Oxford: Oxford University Press.

Hellen, J.A. (1983) 'Primary health care and the epidemiological transition in Nepal', in N.D. McGlashan and J.R. Blunden (eds) *Geographical Aspects of Health*, London: Academic Press, pp. 285–318.

Hewitt, T. (1992) 'Developing countries: 1945 to 1990', in T. Allen and A. Thomas (eds) *Poverty and Development in the 1990s*, Oxford: Oxford University Press, pp. 221–37.

Hewitt, T., Johnson, H. and Wield, D. (eds) (1992) *Industrialization and Development*, Oxford: Oxford University Press.

Hill, A. and Cleland, J. (eds) (1991) *The Health Transition: Methods and Measures*, Canberra: Australian National University.

Institute for Population and Social Research (1988) *Thailand: The Morbidity and Mortality Differentials*, ASEAN Population Programme Phase III Country Study Report, IPSR Publication No. 119, Bangkok: Mahidol University.

Jarrett, S.W. and Ofosu-Amaah, S. (1992) 'Strengthening health services for MCH in Africa: the first four years of the "Bamako Initiative"', *Health Policy and Planning* 7(2): 164–76.

Jones, K. and Moon, G. (1992) 'Medical geography: global perspectives', *Progress in Human Geography* 16(4): 563–72.

Kitching, G. (1982) *Development and Underdevelopment in Historical Perspective*, London: Methuen.

Learmonth, A. (1988) *Disease Ecology*, Oxford: Blackwell.

Leete, R. (1985) 'Increased survival in East and South-east Asia: towards new outer limits to life', in *Proceedings of the International Union for the Scientific Study of Population, International Population Conference, Florence*, pp. 429–42.

Lopez, A. (1989) 'La santé en transition à la réunion de 1946 à 1986', *Annales de Géographie* 546: 152–78.

McGlashan, N.D. (1982) *A West Indies Geographic Pathology Study*, Occasional Paper No. 12, Department of Geography, University of Tasmania.

McKeown, T. (1965) *Medicine in Modern Society*, London: Allen & Unwin.

—— (1988) *The Origins of Human Disease*, Oxford: Blackwell.

Mamdani, M. (1991) *Reproductive Behaviour in Poor Urban Areas: A Review*, Urban Health Programme, London School of Hygiene and Tropical Medicine.

Messkoub, M. (1992) 'Deprivation and structural adjustment', in M. Wuyts, M. Mackintosh and T. Hewitt (eds) *Development Policy and Public Action*, Oxford: Oxford University Press, pp. 175–98.

Olshansky, S.J. and Ault, A.B. (1986) 'The fourth stage of the epidemiologic transition: the age of delayed degenerative diseases', *Milbank Memorial Fund Quarterly* 64(3): 355–91.

Omran, A.R. (1971) 'The epidemiologic transition: a theory of the epidemiology of population change', *Milbank Memorial Fund Quarterly* 49(4[1]): 509–38.

—— (1977) 'Epidemiologic transition in the United States: the health factor in population change', *Population Bulletin* 32(2): 1–42.

Phillips, D.R. (1988) *The Epidemiological Transition in Hong Kong: Changes and Health and Disease since the Nineteenth Century*, Occasional Papers and Monographs No. 75, Centre of Asian Studies, University of Hong Kong.

—— (1990) *Health and Health Care in the Third World*, London: Longman.

—— (1991) 'Problems and potential of researching epidemiological transition: examples from Southeast Asia', *Social Science and Medicine* 33(4): 395–404.

—— (ed.) (1992) *Ageing in East and South-East Asia*, London: Edward Arnold.

Picheral, H. (1989) 'Géographie de la transition épidémiologique', *Annales de Géographie* 546: 129–51.

Richters, A. (1992) Introduction to theme issue on Gender, Health and Development, *Social Science and Medicine* 35(6): 747–51.

Riley, J.C. (1989) *Sickness, Recovery and Death: A History and Forecast of Ill Health*, Iowa City: University of Iowa Press.
Riley, J.C. and Alter, G. (1989) *The Epidemiological Transition and Morbidity*, Working Paper No. 10, Population Institute for Research and Training, Bloomington: Indiana University.
Seers, D. (1969; new edn 1979) 'The meaning of development', in D. Lehmann (ed.) *Development Theory: Four Critical Studies*, London: Frank Cass.
Simmonds, O.G. (1988) *Perspectives on Development and Population Growth in the Third World*, New York: Plenum Press.
Smith, K.R. (1990) 'The risk transition', *International Environmental Affairs* 2(3): 227–51.
Smith, P. (1992) 'Industrialization and environment', in T. Hewitt, H. Johnson and D. Wield (eds) *Industrialization and Development*, Oxford: Oxford University Press, pp. 277–302.
Stephens, C. and Harpham, T. (1991) *Slum Improvement: Health Improvement?*, PHP Departmental Publication No. 1, London School of Hygiene and Tropical Medicine.
—— and —— (1992) 'Health and environment in urban areas of developing countries', *Third World Planning Review* 14(3): 267–82.
Tabibzadeh, I., Rossi-Espagnet, A. and Maxwell, R. (1989) *Spotlight on the Cities: Improving Urban Health in Developing Countries*, Geneva: WHO.
Todaro, M.P. and Stilkind, J. (1981) *City Bias and Rural Neglect: The Dilemma of Urban Development*, New York: Population Council.
Tout, K. (1989) *Ageing in Developing Countries*, Oxford: Oxford University Press.
United Nations Development Programme (1991) *Human Development Report 1991*, Oxford: Oxford University Press.
—— (1992) *Human Development Report 1992*, Oxford: Oxford University Press.
United Nations Children's Fund (1991) *The State of the World's Children 1991*, Oxford: Oxford University Press.
Verbrugge, L. (1984) 'Longer life but worsening health? Trends in health and mortality of middle-aged and older persons', *Milbank Memorial Fund Quarterly* 62(3): 475–519.
Wallace, H.M. and Giri, K. (eds) (1990) *Health Care of Women and Children in Developing Countries*, Oakland, CA: Third Party Publishing.
Weil, D.E.O., Alicbusan, A.P., Wilson, J.F., Reich, M.R. and Bradley, D.J. (1990) *The Impact of Development Policies on Health: A Review of the Literature*, Geneva: WHO.
World Bank (1991) *World Development Report*, New York: Oxford University Press.
—— (1993) *World Development Report*, New York: Oxford University Press.
World Health Organization (1992a) *Our Planet, Our Health: Report of the WHO Commission on Health and Environment*, Geneva: WHO.
—— (1992b) *World Health Statistics Annual 1991*, Geneva: WHO.
Zwi, A. and Ugalde, A. (1989) 'Towards an epidemiology of political violence in the Third World', *Social Science and Medicine* 28(7): 633–42.

2

GLOBAL ENVIRONMENTAL CHANGE AND HEALTH

Graham Bentham

INTRODUCTION

There is a long history of concern about the effects of environmental conditions on health. The slums, polluted water supplies and lack of sanitation of the typical nineteenth-century industrial city have their echo in the appalling living conditions still endured by many inhabitants of the Third World. However, the problems of the Victorian city led to an outcry that had its practical expression in the Sanitary Reform movement. Since then the developed countries of the world have made great strides towards creating more civilized conditions. However, this has led to no lessening of public concern about the environment and health. People still worry about the effects of air pollution from traffic and industry, about nitrates and other contaminants in the water supply and about the possible risks of living near nuclear power plants and hazardous waste dumps. An important common thread in these issues is that the problems are for the most part local in scale. The Victorian industrial city might have been a nightmare, but close by were rural areas with a much less polluted environment to which the more favoured citizens could escape.

The 1980s saw a dramatic new development. For the first time the scientific community and politicians started to take seriously the threat that human activities were damaging the environment on a global scale. A dramatic development was the discovery in 1985 that a hole had appeared in the ozone layer over Antarctica. More gradual was the realization that the build-up of greenhouse gases in the atmosphere threatened to alter the world's climate. The growing recognition of an unwitting ability to alter our environment has been followed by anxiety that the consequences of these changes for human health and welfare could be enormous if steps are not taken to undo the damage. Furthermore, this time the easy option of moving on to escape from the mess that has been made is not a possibility.

OZONE DEPLETION AND ULTRAVIOLET RADIATION

The ozone layer in the stratosphere is a vital shield against ultraviolet radiation from the sun. It absorbs all of the most damaging UVC radiation and much of the UVB radiation that is also capable of harming living things, including people. It is known that the relatively small amounts of shorter-wavelength ultraviolet radiation that do get through are capable of causing diseases such as skin cancer and eye cataracts. Any damage to the ozone shield would allow greater amounts of UVB to penetrate the atmosphere and reach the ground, leading to higher human exposure and a worsening of health impacts. The first scientific proof of such damage came with the publication in 1985 of results showing that large losses of ozone over Antarctica had been occurring each spring since the late 1970s. Intensive monitoring of Antarctic ozone has shown that this ozone hole has recurred seasonally until the present day, with some evidence that the scale of depletion has worsened. Scientific research has shown that the damage is done by reactive chlorine and bromine derived mainly from man-made chlorofluorocarbons (CFCs) and halons. Extreme winter cold leads to the formation of polar stratospheric clouds on the surfaces of which inactive chlorine is converted into forms that can deplete ozone. The persistent winter vortex of westerly winds around the pole is also important in preventing the mixing of the chemically perturbed air with that from lower latitudes.

The Antarctic ozone hole is a spectacular warning of how the natural environment can be damaged by human activities. However, it is a stroke of luck that this depletion has taken place over a part of the world with no permanently settled population to be exposed to the enhanced levels of ultraviolet radiation. This means that the *in situ* health effects of Antarctic ozone depletion have probably been minimal. Of much greater concern is the prospect of ozone losses over populated areas. One way that this can happen is by the transport to lower latitudes of ozone-depleted air following the breakdown of the Antarctic vortex in the spring. The populations of Australia, New Zealand and the southern parts of South America are therefore at risk from ozone depletion and the associated enhanced UVB radiation doses. It may be of little comfort to those affected, but, taking a wider perspective, it is fortunate that relatively few people live at high latitudes in the Southern hemisphere.

Things are very different in the Northern hemisphere, where many millions of people, particularly in Europe, live at relatively high latitudes. Ozone losses in the Arctic would therefore pose much greater problems. However, measurements from the Arctic do not show ozone depletion on a scale similar to the spectacular losses over Antarctica. It appears that the somewhat higher temperatures, lower abundance of polar stratospheric clouds and greater variability of the polar vortex make the Arctic environment less conducive for ozone depletion. Abnormally high concentrations of ozone-destroying chemicals have nevertheless been measured in the Arctic stratosphere, which may be primed for future ozone losses.

Alongside the polar data, some interesting patterns are emerging from

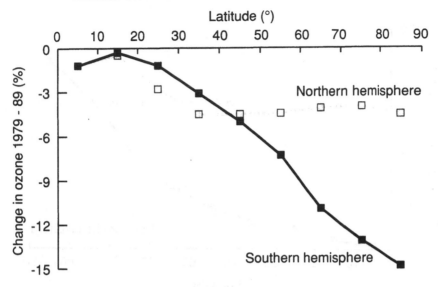

Figure 2.1 Changes in total ozone by latitude, the Northern and Southern
hemispheres, 1979–89
Source: Based on data from Madronich (1992)

studies of ozone trends on a global scale. Ground-based measurements go
back as far as the 1950s, and satellite data first became available in 1978.
Both sets of data show clear evidence of widespread losses of ozone since
about 1970. Furthermore, there are ominous signs that the rate of decline
has accelerated in the last few years. Losses have been greatest for the winter
and early spring and are greater at mid- and high latitudes than they are in
the tropics, with the large depletions being seen in the areas adjacent to the
Antarctic (see Figure 2.1). When the data are analysed by season this reveals
an early spring depletion of about 8 per cent per decade for mid-latitudes
in the Northern hemisphere.

In view of this evidence of ozone depletion on a global scale it might be
expected that human populations would already have been exposed to
enhanced levels of solar ultraviolet radiation. Using the trends in ozone as
an input, it is possible to estimate the amounts of ultraviolet radiation of
different wavelengths reaching the surface. These can then be weighted by
the ability of each wavelength to cause biological damage to produce an
overall estimate of change in biologically effective UV. For example, it is
possible to weight by ability to damage DNA, which is likely to be a
relevant factor in the induction of cancers. Figure 2.2 shows that DNA-
weighted doses over the decade 1979–89 are estimated to have increased by
about 7.5 per cent at 50° north and by about 12.5 per cent at 50° south.

However, these estimates assume constant levels of pollution in the
troposphere, whereas the reality is that this has worsened over this time

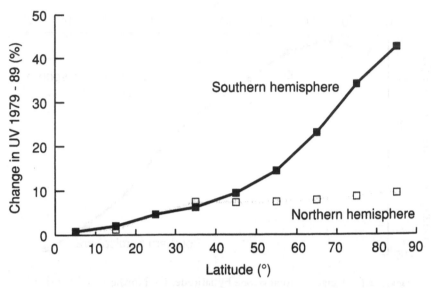

Figure 2.2 Estimated changes in surface UV (weighted by DNA damage) by latitude
in the Northern and Southern hemispheres, 1979–89
Source: Based on data from Madronich (1992)

period. Ironically, some of the worst problems are caused by ozone, which
may be an important UV shield in the stratosphere but is a damaging
pollutant in the troposphere. In some areas these rising levels of pollution
from motor vehicles and industry may have masked some of the potential
increases in UV resulting from stratospheric ozone depletion. This might
account for some surprising results from the United States which show that
UVB levels actually fell between 1974 and 1985, perhaps as a result of
increasing local pollution near the measurement sites. However, high UVB
levels associated with abnormally low levels of ozone have been recorded
for relatively unpolluted areas such as New Zealand. As efforts to control
pollution of the troposphere intensify, it seems likely that this will have the
unwelcome effect of increasing the amount of UV radiation reaching the
ground where it can do harm.

It is clear that the potential is already there for a substantial increase in
the exposure of the human population to ultraviolet radiation as a result of
stratospheric ozone depletion. In spite of current efforts to control the
production of CFCs and halons under the Montreal Protocol, it is expected
that concentrations of these substances in the stratosphere will continue to
rise into the next century. Therefore, there is every prospect of further
ozone depletion which could exacerbate the public health problems posed
by diseases related to exposure to ultraviolet radiation.

Ultraviolet radiation and human health

Non-melanoma skin cancers occur more frequently than any other type of cancer in fair-skinned populations, and their incidence has been rising rapidly for several decades. Just about all the major epidemiological features of the disease point strongly towards exposure to sunlight as a significant cause. They occur most often on the face and neck, which are the parts of the body with the greatest exposure to the sun. They are particularly common in occupations such as farming and fishing where long hours are spent out of doors. A marked latitudinal gradient has been found within and between countries, with risks increasing at lower latitudes where the intensity of ultraviolet radiation is higher. The incidence of non-melanoma skin cancers is particularly high in countries such as Australia and New Zealand where fair-skinned populations have migrated from high to lower latitudes, thereby increasing their exposure to solar radiation. Incidence also shows a marked rise with age, suggesting the importance of cumulative exposure to ultraviolet radiation. The evidence is therefore overwhelming that UV exposure is a major factor in non-melanoma skin cancer.

This being the case, there seems little doubt that higher UV doses as a result of ozone depletion would increase the incidence of these diseases. Estimates of the scale of such increases vary considerably depending on the data and methods used. After a review of the available evidence, the US Environmental Protection Agency has concluded that each 1 per cent decrease in ozone might lead to between 1.5 and 2 per cent additional cases. It estimates that an assumed 10 per cent ozone loss by the year 2050 would lead to about 2 million additional cases of non-melanoma skin cancers in the US population, with even greater numbers for higher levels of depletion. In terms of anxiety for the sufferers and demands on health services this would be a substantial problem. However, the good news is that these cancers can normally be treated successfully, often on an out-patient basis, and they are not usually fatal

Malignant melanoma is an important and rapidly increasing public health problem in fair-skinned populations. As can be seen from Figure 2.3, the number of deaths from melanoma in England and Wales has doubled in the last twenty years, and similar rises have occurred elsewhere. Although in terms of the number of cases it remains much less common than other skin cancers, it affects younger people as well as old, is more difficult to treat successfully and causes more deaths. It shares with non-melanoma skin cancers several of the epidemiological features that point to a causal role for UV exposure. In particular, a latitudinal gradient has been observed within several countries, and the incidence is greatest in countries such as Australia with a fair-skinned migrant population from higher latitudes. It has also been shown that recreational exposure to the sun is significantly greater for melanoma cases than it is for controls. However, other features of its

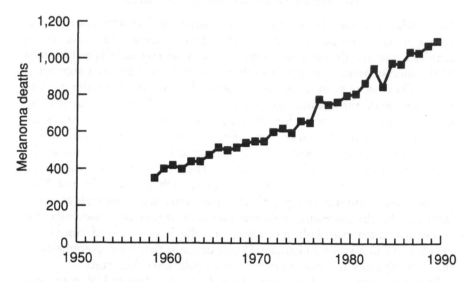

Figure 2.3 Deaths from malignant melanoma in England and Wales, 1958–89

epidemiology show that the risk of melanoma is not simply related to cumulative exposure to sunlight.

The geographical pattern of the disease in Europe shows some marked departures from a simple trend with latitude, with lower incidence in Mediterranean countries than in Scandinavia. Melanoma is not more common in outdoor workers, and it occurs on infrequently exposed parts of the body as well as on commonly exposed sites such as the face and neck. This has led some to cast doubt on the hypothesis that sunlight exposure is an important factor in melanoma. However, this is probably going too far. The discrepancies in the geography of melanoma in Europe can be accounted for by genetically determined differences in skin pigmentation, with the extremely fair-skinned Scandinavians being at especial risk. The occupational, anatomical and age distribution of the disease could be accounted for by the hypothesis that intermittent, intense exposure on unacclimatized skin is most important. This underlines the importance of factors such as clothing styles, patterns of outdoor recreation, sunbathing habits and the trend towards foreign holidays. On balance it is prudent to assume that exposure to ultraviolet radiation is a factor in malignant melanoma and that risks will be increased as a result of ozone depletion.

Reflecting these greater complexities, the US Environmental Protection Agency's risk estimates show a wider range of uncertainties for melanoma than for non-melanoma skin cancers. Its estimate is that a 1 per cent decrease in ozone might be associated with an increase of between 1 and 2 per cent in the number of cases of malignant melanoma. For an assumed depletion of ozone of 10 per cent by the year 2050 this could mean an additional 4,000 melanoma deaths in the US population, while 50 per cent ozone depletion would be expected to produce 18,000 additional deaths.

Any impacts of ozone depletion on skin cancers would vary greatly between different populations. These cancers are rare in people with moderately or highly pigmented skins, and any effects of ozone depletion on incidence would be largely confined to the minority of the world's population with fair skins. However, this would not be the case for some other possible health impacts. For example, blindness as a result of the formation of cataracts is common in all populations and is a major source of untreated disability in many Third World countries. There have been suggestions that the risk of developing some types of cataract is increased by exposure to sunlight. For example, the prevalence of cataracts in rural Aborigines in Australia has been shown to be related to UV levels in different areas.

Cataracts have also been shown to be more common in the southern than the northern states of the USA, which is in accord with differences in ultraviolet radiation. However, doubts have been expressed as to whether it is UV that is the causal factor. It has been pointed out that cataracts are more common in poorer social groups, who also tend to suffer more from diarrhoeal diseases, which may cause cataracts. Traditionally, social conditions and diarrhoeal diseases have been worse in the southern states, and it may be this, rather than their levels of UV radiation, that is the explanation for their excess of cataracts. However, there is support for the UV hypothesis from a study of cataracts in a group of watermen working on Chesapeake Bay in the United States. This found a significant positive relationship between individual levels of UV exposure and cortical cataracts (but not other types) which is unlikely to have been the result of confounding by other factors. It therefore seems likely that some cataracts are caused by sunlight exposure, but quantitative assessments of the possible impacts of ozone depletion must await research to disentangle UV effects from other factors.

Another effect that would be no respecter of race or colour would be the possible impacts of enhanced ultraviolet radiation doses on infectious diseases. Research has shown that UV can activate the human immunodeficiency virus (HIV) which could hasten the onset of AIDS in infected individuals. It is also known that UV can affect the human immune system, which means that enhanced exposure might reduce resistance and increase the spread of some infectious diseases. However, direct evidence of such impacts on human populations is presently lacking.

In addition to these potential direct impacts of enhanced UV radiation it is possible that human health might be affected in more indirect ways. Some agriculturally important plants are sensitive to UV radiation, and yields might be reduced as a result of enhanced exposure. There might also be adverse effects on marine phytoplankton organisms, which are at the base of some important food chains. Therefore, ozone depletion might lead to changes in the availability and price of food supplies, which could worsen the nutritional status and health of vulnerable groups.

If the worst comes to the worst, the impacts of ozone depletion on the health of the world's population could be very great. Continued increases in the concentrations of CFCs and halons in the stratosphere could accelerate the destruction of ozone, leading to substantially enhanced exposure to ultraviolet radiation. There seems little doubt that, other things being equal, this would increase greatly the risks of skin cancers. However, the impacts could turn out to be much wider than this if the fears about cataracts and infectious diseases and of indirect effects of UV exposure turn out to be true. However, much of this is uncertain, and it could be that the main impacts will be confined to increasing skin cancer risks in the fair-skinned populations of the richer countries of the world. In terms of numbers affected, the biggest impact would be likely to be on non-melanoma skin cancers, which are usually treatable and rarely fatal. The populations at greatest risk would also be those with generally good access to medical care. Malignant melanoma carries a higher risk of death, but the numbers involved are relatively small. Furthermore, changes in behaviour involving simple measures such as wearing a hat, using sun-screens and avoiding the sun in the middle of the day could do much to reduce risk.

The health risks posed by stratospheric ozone depletion could therefore turn out to be less than catastrophic. In particular, it seems unlikely that they will be as widespread or severe as some of the health problems that might be associated with global warming as a result of the enhanced greenhouse effect.

GLOBAL WARMING

Greenhouse gases are responsible for trapping infra-red radiation in the atmosphere and so warming the surface of the Earth. Without them the temperature would be about 33°C lower, and life would be impossible. Therefore, the natural greenhouse effect is a benign influence on which the habitability of the planet depends. The problem is that human activities are adding to the concentrations of the major greenhouse gases (carbon dioxide, CFCs, methane and nitrous oxide) in the atmosphere, leading to an enhancement of the greenhouse effect and a consequent warming of the global climate. There are already signs of this happening in the instrumental record of global mean surface air temperatures, which have risen during the nineteenth and twentieth centuries. Furthermore, the evidence of warming is particularly striking for the 1980s, although it can be difficult to disentangle the effects of an enhanced greenhouse effect from natural climatic variability.

As far as future changes are concerned, the most authoritative assessment has been made by the Intergovernmental Panel on Climate Change (IPCC). If it is assumed that the present trend in emissions of greenhouse gases continues into the future, the IPCC predicts that global mean temperatures will increase by about 0.3°C per decade. By the end of the next century this would lead to temperatures about 4°C above the pre-industrial average. More recent studies have tended to reduce these estimates somewhat. Nevertheless, the scale of the anticipated change would still be greater than in any period of human history and would approach in magnitude the changes that took place at the end of the last Ice Age.

Given the interdependence of different aspects of the climate system, global warming will also lead to major changes in precipitation, cloud cover and the frequency, intensity and location of storms. Regional variations in the effects of this greenhouse gas forcing will mean that some areas will experience changes in climate even more dramatic than is suggested by global averages. However, reliable predictions on a detailed regional scale, particularly for variables such as cloud cover and precipitation, are currently lacking. Substantial changes can also be expected for other parts of the natural environment that are dependent on climate. Alterations to temperatures, cloud cover and precipitation will affect soil moisture, which is of vital importance for natural ecosystems and for agriculture. Mean sea-level is also expected to rise by about 6 cm per decade as a result of thermal expansion of ocean water and the melting of some land ice.

It is important to stress that the significance of such developments lies not only in the magnitude but also in the rate of changes. The recovery of the climate at the end of the last Ice Age took several millennia, whereas substantial global warming is anticipated within the lifetime of children alive today. It is this pace of change that will add enormously to the difficulties that are posed by these events.

The health effects of global climate change

If the climate does change in the way that is being predicted, there are likely to be enormous implications, not least for human health. Precise prediction of what these might be is obviously very difficult. One stumbling block is that future global trends in important climatic variables such as temperature and precipitation are themselves very unclear, and these uncertainties become much greater when predictions have to be made for particular regions. Then there are immense problems in predicting what effects these changes might have on health. Here much reliance has to be placed on studies of the links between disease and seasonal and geographical variations in the important climatic variables. The hope is that these might provide analogues for the sort of changes that are predicted for the future. This type of approach is likely to be most successful where the effects of weather and

41

climate on health are fairly direct, as might be the case with physiological reactions to extremes of heat and cold. However, many of the impacts of climate change might not be like this. Instead, they are likely to involve complex interactions between many factors, as when climate change leads to alterations in natural ecosystems which in turn affect animals and insects that are important vectors for human disease.

The complexities become immense when it is realized that the consequences for health will also be profoundly affected by the socio-economic circumstances of the communities that are involved. It needs to be recognized that many of the impacts of climate change will be on the vulnerable populations of the Third World who already face enormous problems in maintaining good health in their existing physical and socio-economic environment. Furthermore, the environmental changes that will unfold in the next century will be set in a context of an expanding population placing increased pressure on the resources that are available to support it. In view of these difficulties it is possible only to give pointers to some of the health impacts that might be anticipated. It is useful to begin by discussing the more direct effects, before moving on to consider some of the more indirect and complicated ways in which climate change might affect human health.

The most obvious and direct effect of global warming might be the increased morbidity and mortality associated with episodes of extreme high temperature. As well as the effects of heat stress, adverse effects on cardiovascular health are likely, especially in vulnerable groups such as the elderly. An analysis of coronary heart disease and stroke mortality for a sample of US cities has shown that mortality rates tend to rise sharply once daily

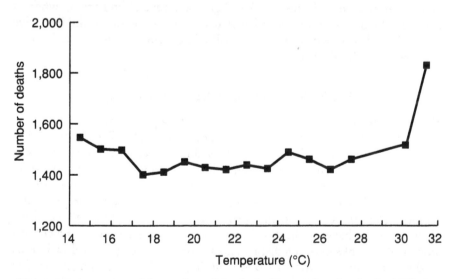

Figure 2.4 Average weekly number of deaths in Greater London in the 1970s by weekly average of maximum daily temperatures

temperatures exceed about 25°C. Mortality rates in Britain during the exceptionally hot summer of 1976 also showed a marked rise during the hottest period, illustrated by data for the 1970s in Figure 2.4.

If the greenhouse effect leads to a rise in average temperatures while the extent of day-to-day variability remains constant, the frequency of such episodes is likely to increase sharply. Table 2.1 shows how predicted changes in summer temperature in England would affect the probability of occurrence of a summer as hot as that of 1976. On the basis of normal distribution theory and data for the historical record of summer temperatures, the probability of occurrence of a summer as hot as 1976 is about 0.1 per cent. However, with a predicted rise in temperature of 1.4°C by the year 2030 there would be a hundredfold increase in the probability of such a hot summer. By 2050, when temperatures are predicted to have risen by 2.1°C, summers as hot as 1976 can be expected to occur in one out of three years. This underlines the fact that what appear to be relatively modest changes in averages can produce dramatic effects on the rate of occurrence of extreme events.

Table 2.1 Effect of predicted climate changes on the probability of occurrence of a summer as hot as 1976 in central England

Year	Predicted temperature change (°C)	Predicted temperature (°C)	Probability
1991		15.3	0.001
2010	+0.7	16.0	0.018
2030	+1.4	16.7	0.011
2050	+2.1	17.4	0.355

Source: based on data from Warrick and Barrow (1991)

Food poisoning is another health problem that shows a strong positive association with temperature, reflecting the increased reproduction rates of bacteria in warmer conditions. Depending on the micro-organisms involved, the doses received and the susceptibility of the affected individual, effects can range from the trivial to the life-threatening. Since many of the less serious cases of food poisoning are not reported, it is not possible to be precise about the true extent of morbidity from this cause. Nevertheless, it is clear that it is an important cause of illness, with major economic costs even in countries such as the United Kingdom and the United States with generally good standards of hygiene.

Brutally direct effects of weather and climate on health can also be seen in the injury, disease and loss of life associated with storms and floods. These events can occur in many parts of the world, but undoubtedly the most severe effects are associated with tropical storms (typhoons, hurricanes or cyclones). The south Florida hurricane of 1992 was a reminder of the immense damage that can be associated with these events, even in a rich country with well-developed infrastructure. In poorer countries they can cause immense human suffering, as was the case in the Bangladesh cyclone of 1988. Since such storms form only where sea surface temperatures are 27°C or greater, a warmer climate could therefore lead to them affecting a larger area and for a greater part of the year. Their maximum intensity, which is dependent on temperature, could also increase, adding to their potential for destruction. In low-lying coastal areas the risks of flooding associated with such storms would be further enhanced by any rise in mean sea-level resulting from global warming.

The effects of such events can go far beyond the immediate loss of life and injuries associated with the storms and floods themselves. Disruption and contamination of water supplies can be widespread, increasing the potential for epidemics of water-related diseases. Agricultural production can be lost, leading to a decline in local food availability, a rise in prices and a decreased ability of the local population to buy food because of their loss of income. Malnutrition is therefore likely to become more widespread and severe, decreasing the resistance of the population to infections. Such problems are likely to be exacerbated where refugees are crowded together in insanitary conditions. However, it is important to recognize that such devastation is a reflection not only of the severity of the storm but also of the vulnerability of the local population. Where tropical storms affect an affluent population with good general infrastructure and efficient systems for warning, shelter and evacuation, it is possible for the health impacts to be minimized.

In temperate countries the health effects of warmer conditions would by no means all be bad. Mortality rates, particularly from cardio-vascular and respiratory diseases, tend to rise during colder weather, and once again it is the vulnerable members of the population such as the elderly and infants who are most at risk. In Britain it has been estimated that each 1°C by which the winter is colder than average is associated with 8,000 excess winter deaths. Warmer winters might therefore be expected to lead to a reduction in mortality rates. However, the actual outcome may well be different from this. It has been argued that part of the rise in mortality during colder weather is the result of the inability of some people to afford adequate domestic heating, a situation that has been described as fuel poverty. This could be exacerbated if, as seems likely, governments respond to the threat of global warming by increasing taxes on fuel in order to curb carbon dioxide emissions. There could therefore be the paradoxical situation that mortality related to fuel poverty could rise in spite of warmer winters.

Another impact that is likely to be particularly prominent in some temperate countries is the exacerbation of photochemical air pollution. Photochemical reactions involving nitrogen oxides and volatile organic

44

compounds (principally from motor vehicles) in the presence of sunlight are responsible for the formation of tropospheric ozone and photochemical smog (WHO 1990). In temperate climates the likelihood of enhanced levels of ozone is greatest in the summer months during anticyclonic conditions when winds are light, temperatures are high and there is strong sunlight. Global warming, together with enhanced ultraviolet radiation flux as a result of stratospheric ozone depletion, is likely to encourage the formation of such pollution. Photochemical pollution episodes could begin to afflict areas where present climatic conditions make them rare, while the severity and frequency of smog could also increase in areas that are already regularly affected. Ozone at concentrations measured in ambient air has been shown to cause dose-dependent changes in lung function, with effects that may be particularly troublesome for asthmatics and others with compromised lung function. Exposure to ozone has also been shown to increase the sensitivity of asthmatics to other pollutants, which is a problem because high concentrations of ozone often coincide with high levels of other pollutants.

Vector-borne diseases such as malaria, schistosomiasis, dengue fever and yellow fever have ravaged humanity throughout history. For the most part these diseases have been brought under control in the industrialized countries, but they continue to be a huge problem in the Third World. Climatic factors are often very important influences on the geographical spread and seasonal incidence of such diseases. The rate of development of the pathogen or parasite itself can be temperature-dependent. Similarly, reproduction, longevity and activity of vectors such as mosquitoes can be strongly affected by temperature and precipitation. More indirectly, different species are adapted to different ecosystems, which are themselves affected by climatic conditions. Furthermore, the spread of disease is often linked to human activity patterns, which can be climate-related, as when people are forced to congregate around remaining sources of water during droughts. There is therefore considerable potential for climate change to bring about significant changes in the distribution of such diseases (see Table 2.2).

Regions at higher latitudes or altitudes may become more favourable for the transmission of disease as temperatures rise. There may also be shifts from seasonal to perennial transmission in areas where cold season conditions marginally inhibit the disease. Areas bordering currently affected areas and sharing similar socio-economic circumstances are obviously at risk, and there is a need for a monitoring of conditions in such areas. There has also been concern about the spread of vector-borne diseases into richer temperate countries. Malaria was once common in parts of Western Europe and North America, and epidemics of yellow fever occasionally occurred in parts of Southern Europe and the United States. Stringent control measures, changes in lifestyles and improved socio-economic conditions have combined to eliminate such diseases from these areas except for sporadic outbreaks

Table 2.2 WHO estimates of the major vector-borne diseases likely to be affected by climate change

Disease	Prevalence of infection (millions)	Effects of climate change on distribution
Malaria	270	Highly likely
Schistosomiasis	200	Very likely
Dengue fever	(30–60 million infections p.a.)	Very likely
Lymphatic filariases	90	Likely
Onchocerciasis	18	Likely
African trypanosomiasis	(25,000 new cases p.a.)	Likely
Yellow fever	>3,000 deaths in 1986–8	Likely
Japanese encephalitis	no estimate available	Likely

Source: based on WHO (1990)

related to imported cases. It seems likely that these will continue to be limiting factors even in the face of climate changes that enhance the potential for the spread of such diseases.

Another major public health success of the industrialized world has been the control of water-related diseases. However, in many Third World countries where inadequate sanitary conditions prevail and where water quality is poor, water-borne diseases such as typhoid, cholera and childhood diarrhoea continue to exact a heavy toll. Water-washed diseases related to poor personal and food hygiene are also a major problem where adequate quantities of water are lacking. Because predictions of future trends in temperature and especially rainfall for local areas are so uncertain, it is impossible to make precise predictions of what effects they might have on water quality and quantity. In some areas increases in rainfall might improve the availability of water. Elsewhere, higher temperatures or reduced rainfall may exacerbate water shortages and increase the reliance on poorer-quality sources of supply, thereby increasing the risks of both water-borne and water-washed disease. A particular problem is likely to be faced by areas dependent on mountain meltwater for spring and summer supplies. In these regions higher winter temperatures could lead to more precipitation falling as rain than as snow, thereby reducing the availability of meltwater when it is needed. Another distinctive problem will be in low-lying areas where sea-level rise leads to increased salinity of water supplies.

Of all human activities, farming is the most intimately related to environmental conditions, and therefore climate change seems destined to have major impacts on agriculture. Some areas can expect an increase in production where higher temperatures or increased precipitation lead to a relaxation of present climatic constraints on agricultural productivity. There are also suggestions that greater temperature gradients between continents and oceans will increase the intensity of the monsoons on which many millions of people depend. Elsewhere, decreased water availability during hotter summers, as is predicted for major cereal-growing areas in North America and Western Europe, could reduce agricultural output. Climate change could also bring problems of increased storm damage, greater weed growth and more favourable conditions for the spread of agricultural pests and pathogens.

The worst effects are likely to be in areas where the balance between people and food is already precarious because of population pressure, a marginal climate or existing problems of environmental deterioration, as in much of sub-Saharan Africa. Serious difficulties are also possible in densely populated agricultural areas such as Bangladesh and the Nile delta of Egypt that are vulnerable to salinization or flooding as a result of sea-level rise. In similarly vulnerable areas of richer countries, such as the Netherlands, technology and resources are likely to be adequate to meet this challenge by improved coastal protection. However, in the poorer countries there are nightmare prospects of sea-level rise leading to major losses of agricultural production, loss of livelihood for millions of people and the creation of unprecedented numbers of environmental refugees. In such circumstances the risks of famine can only increase as a result of a decline in local food availability and a reduction in food entitlement because of the inability of those who have lost their livelihoods to purchase imported food. In richer countries any reductions in local agricultural output will simply be made good by imports, albeit at a cost. However, adverse effects on production in the major food-surplus areas of Europe and North America would reduce world food stocks and increase prices. This could have devastating consequences for those countries that rely on food aid to make up for local deficiencies of production. Even where outright famine is avoided, the threat of an increase in chronic malnutrition will remain in large parts of the world.

CONCLUSIONS

There has been no shortage of sensationalist stories of the threats posed by global environmental change, and the public can be forgiven for its scepticism about the more overblown claims that have been made. However, there is also a danger of complacency, which could stand in the way of the actions that need to be taken to counter the real threats that do exist. It is worth repeating that the existence of a recurrent ozone hole over Antarctica is an established scientific fact. The scientific evidence of a continuing build-up of ozone-destroying chemicals in the stratosphere is also compelling, as is that for a gradual but accelerating global depletion of ozone. Nor is there any doubt about the increasing concentrations of greenhouse gases in the Earth's atmosphere. When attention turns to what might happen in the

47

future, inevitably the uncertainties become greater, and it becomes easy by careful selection of assumptions to cast doubt on the accuracy of the predictions that have been made. However, the atmospheric scientist Stephen Schneider has warned that the sword of uncertainty has two blades, and current models may just as well underestimate as overestimate future changes.

The Montreal Protocol on the control of production of CFCs and halons bears witness to the seriousness with which the world's politicians view the threat to the ozone layer. No such political consensus exists on global warming, but the considered view of the world's leading atmospheric scientists is that there is a real prospect of an unprecedentedly rapid change in climate. Although great uncertainties still remain, a prudent conclusion is that there is a substantial risk of changes to the global environment that could have enormous implications for human activities and welfare, not least for human health.

Some of the health impacts of future environmental changes are likely to be fairly direct, which makes prediction somewhat easier. For example, exposure to ultraviolet radiation is a dominant factor in the aetiology of skin cancer, and therefore it can be predicted with some confidence that enhanced exposure to UV will increase risk. Other impacts on health will be the result of much more complex processes, as is the case with climate-related changes in vector-borne diseases and water and food availability. For these problems climate change will be only one factor among many, and predictions will need to take into account this multiplicity of causal factors. Inevitably, this will mean that assessments of these impacts will be subject to much greater degrees of uncertainty. In such cases it will be vital to take into account the socio-economic circumstances of the populations experiencing climate change, since this will profoundly affect the resulting health consequences. There can be no more dramatic example of this than the case of sea-level rise. To the typical citizen of the Netherlands this may mean little more than marginally higher taxes to pay for coastal protection, whereas a poor farmer in Bangladesh may face a serious risk of drowning, loss of livelihood and starvation. More generally, it is clear that the health impacts of global environmental change will not fall equally on the world's population. Lacking the necessary infrastructure and financial resources, it will be poorer countries that will find it most difficult to respond effectively to problems posed by climate change. Within countries it will be the poor and powerless and those already weakened by disease who will be most affected.

Some critics have sought to minimize the health threats posed by global environmental change. By discounting some of the less certain effects it is possible to argue that the health impacts of ozone depletion may amount largely to an increase in mostly treatable skin cancer morbidity. Even for the more wide-ranging effects of global warming, it can be argued that these are likely to be less than the health impacts of existing problems of poverty and insanitary living conditions in the Third World and the future threat of a continued growth of world population. This may well be true. However, this is a reflection of the overwhelming magnitude of these problems, not of the insignificance of the menace posed by global environmental change. Also, such arguments miss the key point that the problems resulting from environmental change will interact with and add to these other threats to human welfare. Many of the same unfortunate people will be in the front line of all these battles.

REFERENCES AND FURTHER READING

Haines, A. and Fuchs, C. (1991) 'Potential impacts on health of atmospheric change', *Journal of Public Health Medicine* 13: 69–80.

Leaf, A. (1989) 'Potential health effects of global climatic and environmental changes', *New England Journal of Medicine* 321: 1577–83.

Leggett, J. (1990) *Global Warming: The Greenpeace Report*, Oxford: Oxford University Press.

Madronich, S. (1992) 'Implications of recent total atmospheric ozone measurements for biologically active ultraviolet radiation reaching the Earth's surface', *Geophysical Research Letters* 19: 37–40.

Russell-Jones, R. and Wigley, T. (1989) *Ozone Depletion: Health and Environmental Consequences*, Chichester: Wiley.

Warrick, R.A. and Barrow, E.M. (1991) 'Climate change scenarios for the UK', *Transactions of the Institute of British Geographers*, n.s. 16(4): 387–99.

World Health Organization (1990) *Potential Health Effects of Climatic Change*, Geneva: WHO.

3

ECONOMIC CRISIS, ADJUSTMENT AND THE IMPACT ON HEALTH

Sheena Asthana

INTRODUCTION

The post-war period has witnessed a number of paradigm shifts in the health and development debate. For the most part, approaches to health have shared the assumptions of the dominant development models of their time. In the 1950s and 1960s, for example, economic development strategies were based upon the belief that less developed countries would repeat the experience of industrialized countries. Health policies similarly followed the health care model of the Western world, medical services being mostly curative, urban-based and highly technological.

In the late 1960s and early 1970s it became clear that economic growth was not 'trickling down' to the poorest sections of Third World societies, and economic theorists began to question the practice of defining development in terms of national income. Dudley Seers, for example, argued that it was more pertinent to look at unemployment, inequality and poverty (1969); if these had all worsened, then even if gross national product was rising, this was hardly a measure of development. Seers suggested that economic growth policies had to be accompanied by redistribution of wealth. In the 1970s a number of international agencies supported this view, and the health focus shifted in favour of the poor, with an emphasis on basic needs.

The late 1960s also saw the emergence of a radical theory of development, the *dependency approach*. This viewed underdevelopment not as a phase which pre-dated economic growth but as a historical condition of Western capitalism. Dependency theorists argued that, in order to escape from the disadvantages of peripheral status, Third World countries had to cut off all links with industrialized nations and develop autonomously. China, Cuba, Mozambique and other countries which had adopted a broadly socialist development path were held up as successful examples of the way forward for less developed countries.

The dependency school was to have a very strong impact on the health

50

field. The World Health Organization and UNICEF (the United Nations Children's Fund) openly expressed enthusiam for the Chinese strategy of development and incorporated many elements of the Chinese health model into their statement on primary health care (PHC). PHC ideals were widely accepted, bringing about another paradigm shift in the health debate. Today it is common for observers on both the left and *right* of the political spectrum to acknowledge that the health problems of the Third World poor stem less from an *absolute* lack of resources than from the massive disparities that exist between rich and poor.

Despite this consensus, since the late 1970s health and development policies diverged. At the same time as the architects of PHC were calling for a fundamental redistribution in power and resources, most developing countries began to implement market-oriented economic policies prescribed by international financial institutions. These policies, which are collectively known as *economic adjustment*, have been imposed on indebted Third World countries as conditions of further financial assistance. However, there is plentiful evidence to suggest that they have contributed to an unprecedented decline in the health and living standards of the Third World poor.

This chapter begins by exploring the background to the economic crisis in the Third World in order to understand why adjustment programmes have been so widely implemented. After briefly describing the nature and rationale of adjustment packages, the impacts of adjustment on nutrition, urbanization and health expenditure are addressed. Reference is made throughout this section to the negative impact of adjustment on the health and welfare of women. The chapter concludes by considering alternative solutions to the economic crisis.

BACKGROUND TO THE ECONOMIC CRISIS

While the 1950s and 1960s were a period of steady economic growth in most Third World countries, the modernization strategies adopted by newly independent governments locked their economies into the unequal trading relationships that lie at the root of the current economic crisis. Despite efforts to industrialize and diversify, many less developed countries (LDCs), remained dependent on the export of primary commodities to developed countries for foreign exchange earnings. Because of adverse terms of trade, primary exports generated insufficient wealth to import manufactured goods, industrial supplies and agricultural technology. Third World governments were therefore forced to rely on foreign investment and foreign loans to finance economic development. In what was to become a vicious cycle, such borrowing increased the need to earn foreign exchange to repay loans and hence the need for primary produce.

Primary commodity dependency combined with external debt was to prove disastrous for many low-income nations. In 1973 and 1979 the Organization of Petroleum Exporting Countries (OPEC) increased the price of oil, which led in turn to a rise in the price of manufactured goods. The import bills of LDCs, the vast majority of which were oil importers, rose substantially. At the same time, commercial banks were looking for a profitable way to invest the surplus of petrodollars earned by OPEC countries. Rather than cutting their volume of imports, oil-importing nations took advantage of cheap overseas loans.

Because real interest rates were low during the 1970s, LDCs could rely on inflation to erode the value of their debts (Woodward 1992). The recession of the late 1970s and 1980s, however, had serious consequences for Third World borrowers and their debt repayments. The depressed markets of the United States and other industrialized nations combined with their anti-inflationary policies reduced the demand for primary products and led to a dramatic fall in commodity prices. At the same time, growing trade protectionism in the West increased the price that Third World countries had to pay for imports. The impact of declining terms of trade was compounded in the early 1980s when the United States raised its interest rates in an attempt to attract foreign capital. International interest rates soared, massively increasing the costs to borrowing countries of servicing existing debts.

As a result of the growing debt crisis, the 1980s became a development disaster, particularly for the heavily borrowing countries of Latin America and the primary-commodity-dependent nations of sub-Saharan Africa. According to World Bank figures, between 1980 and 1989 the debt of LDCs increased from $580 billion to $1,341 billion, and it now represents 41.2 per cent of their combined gross national product (GNP). By the early 1980s Third World countries were paying more in interest and repayments on loans than they were receiving in new finance. According to World Bank figures, a net total of $223 billion was transferred from the poorer countries of the Southern hemisphere to the financial institutions of the North between 1983 and 1989. On average, debt service obligations claim 22 per cent of the developing world's export revenues, and 31 per cent in the highly indebted countries of Latin America and the Caribbean. Due to the combined impacts of debt and inflation, this region has witnessed severe economic decline. Average income per capita fell by 7 per cent between 1980 and 1988 (by 16 per cent if account is taken of deteriorating terms of trade). Net investment per capita fell 50 per cent between 1980 and 1985, and unemployment grew by more than 6 per cent a year (UNDP 1990).

In sub-Saharan Africa (SSA), the situation is even more bleak. Many countries in the region continue to rely on one or two agricultural commodities for the bulk of their export earnings. As the real prices for these commodities dropped by more than 40 per cent through the 1980s, SSA's terms of trade have declined significantly (Hewitt 1992). Other factors exacerbating poverty in the region are wars, drought, environmental deterioration and the growing numbers of refugees. Defence imports, for example, absorbed a large amount of foreign exchange in the war-torn countries of Mozambique, Angola and Ethiopia. In Tanzania, Uganda, Zambia and Zaire, proximity to conflict and civil unrest have also promoted increased military spending.

Economic crisis is therefore particularly acute in sub-Saharan Africa. By 1989 the region owed 96.9 per cent of its annual wealth, compared to 26.8 per cent in 1980. During the same period, per capita incomes declined by 25–30 per cent. With an annual increase in unemployment of 10 per cent between 1980 and 1985, and a fall in real wages, both absolute and relative levels of poverty have increased. According to World Bank estimates for 1989, 52 per cent of the region's population are 'poor', and 35 per cent are 'extremely poor'.

THE RESPONSE TO THE DEBT CRISIS: ECONOMIC ADJUSTMENT

As the debt crisis deepened in the early 1980s, many Third World countries suspended part or all of their debt service payments. In 1982 a large debtor, Mexico, announced that it could not afford to meet its immediate obligations. Losing confidence in the security of international lending, the commercial banks became much more reluctant to provide new loans (Woodward 1992). Unable to secure the necessary funds to sustain their economies, Third World debtors had little choice but to turn to the World Bank and the International Monetary Fund (IMF). The latter, however, would only negotiate the rescheduling of loans and the provision of new finance on condition that debtor nations agreed to implement economic adjustment policies. These are designed to reduce imbalances in the economy, both on external accounts and in domestic resource use (Cornia *et al.* 1987).

Economic adjustment is based upon the belief that growth can only be achieved through the operation of the free market. Adjustment packages typically include both short- and long-term measures. *Stabilization programmes*, which are identified closely but not exclusively with the IMF, aim to produce an immediate effect. They involve the correction of balance of payment imbalances and the lowering of inflation through currency devaluation, import reduction and the implementation of a range of austerity measures. The latter, which are designed to reduce budget deficits and to lower the demand for external resources, include the removal of state subsidies and the reduction of public expenditure (especially in non-productive sectors).

The World Bank's *structural adjustment programmes* (SAPs) are designed to promote economic change in the longer term. Focusing less on import reduction than on the need to encourage exports, SAPs include trade liberalization through the removal of import controls and export taxes; the promotion of private foreign investment; the strengthening of export sectors with a comparative advantage (i.e. primary commodities); the privatization of government industries; and market deregulation (for example, the removal of subsidies and relaxation of government regulations such as minimum wage legislation). The underlying principle of these measures is to increase the role of the private sector and to reduce that of the public sector, thereby making the economy more market-oriented (Woodward 1992).

Since the late 1970s, adjustment policies have been widely adopted by Third World governments. There is, however, strong opposition to this market-oriented strategy. Some critics suggest that the IMF and World Bank are acting mainly in the interests of Western financial institutions. The security and profits of the latter are maintained by the fact that adjustment ensures that debt service is extracted from Third World nations. The financial crash that threatened Western banks should large debtors have succeeded in defaulting on their payments has therefore been averted (Kanji et al. 1991).

A second criticism that has been levelled at the adjustment model is that, by focusing on the economies of individual countries, it ignores the fact that the economic problems faced by LDCs stem in large part from world market tendencies outside their control. Indeed, there is little evidence to suggest that adjustment can have a beneficial impact on structurally weak economies. In practice, stabilization policies cause a sharp decline in national income, and the longer-term structural adjustment measures reinforce the unequal trading relationships that led to impoverishment in the first place. A serious problem arises from the assumption that underdeveloped countries have a relative advantage in the production of primary exports. As more and more countries increase their exports of a limited number of primary commodities, the resulting increase in global supply causes commodity prices to plummet.

Perhaps the most fundamental objection to structural adjustment relates to its impact on health and welfare. As critics of earlier modernization strategies pointed out, the benefits of market-oriented economic growth rarely 'trickle down' to the poorest sections of society. Indeed, there is powerful evidence to suggest that the key to understanding patterns of ill-health lies in the distribution of resources within a country rather than the absolute levels of economic growth. In the 1970s the examples of Kerala State in India and China were used to demonstrate how low-income countries can achieve impressively high levels of health through a political commitment to equitable wealth distribution. Such evidence played a significant role in the birth of the primary health care (PHC) movement, supporters of which suggest that, in countries where power and resources are highly concentrated in the hands of the rich, levels of health are abysmally low (see Chapter 12). As in the long term market-oriented strategies tend to result in a skewed income distribution in which the poor have a small share of a nation's wealth, critics of adjustment doubt that it will have a positive effect on health and welfare.

ECONOMIC CRISIS AND ADJUSTMENT: THE IMPACT ON HEALTH

If the long-term prospects of adjusting countries are subject to considerable debate, few deny that the impact of adjustment has been negative in the short term. Rising unemployment and the fall in real wages, the promotion of the cash-cropping sector at the expense of landless labourers and urban dwellers, and the removal of food subsidies have led to a deterioration in living standards, reflected in the rise in malnutrition and infectious disease. At the same time, cuts in public expenditure and the emphasis on cost recovery have made vital public services increasingly inaccessible to those who have the greatest need for them. Adjustment has been particularly hard on the urban poor, whose plight has received growing attention over the past decade (see Chapter 7). However, the suggestion that the rural poor have positively benefited from SAPs has been challenged with evidence that urban migration has increased during the 1980s. In sub-Saharan Africa, this has resulted in growing numbers of female migrants who, in the absence of other employment opportunities, turn to prostitution. The greatest burden of adjustment, it is commonly argued, falls on to women. Their contribution to the household economy has become critical under economic crisis, and increasing numbers are seeking paid employment. At the same time, because of cuts in state service expenditure and the rise in food prices, women are having to spend more time in their role as household managers.

The following sections explore in more detail the impact of economic adjustment on health and welfare. However, two factors make such an assessment difficult. The first is that one cannot easily separate out the effects of debt and the effects of economic adjustment policies. The second is that the impact of economic changes on health is, to a great extent, cumulative over time. This means that the full impact of adjustment will only become apparent over a long period (Woodward 1992).

Adjustment, nutrition and health

In contrast to the relatively high levels of chronic and degenerative diseases that are found in the West, Third World health profiles continue to be dominated by high infant and child mortality from infectious diseases such as diarrhoea, measles and acute respiratory infections. The high incidence of infectious diseases in LDCs cannot be properly understood without considering the role of nutrition. A low nutritional status predisposes an individual to infection, and, because the immunological status of a malnourished individual is impaired, the course of infection is more severe than in a well-nourished person.

The relationship between malnutrition and infection is *synergistic*. Not only are malnourished people more likely to contract and to die from infectious diseases, but the infection itself causes further malnutrition. According to UNICEF, all infections have a nutritional impact. They can depress the appetite, decrease the body's absorption of nutrients, induce rejection of food by vomiting or drain away nutrients through diarrhoea (UNICEF 1983). As a result, malnutrition and infection become joined in a self-reinforcing cycle, and, rather than being disease-specific, most child deaths are caused by a combination of poor nutrition and repeated bouts of different infections.

Because of the critical role of nutrition, it is important to examine the effects of adjustment on food security at household levels. Advocates of adjustment argue that the shift towards primary commodity producers positively benefits rural dwellers, thereby redressing the rural–urban disparities in wealth. Evidence suggests that rural producers with sufficient land for surplus production and income to pay for necessary agricultural and labour inputs have benefited from the rise in producer prices and improved rural–urban terms of trade. However, landless rural labourers who have to buy their food on the open market and smallholders who are net food purchasers have been badly hit by reductions in food subsidies and rising food prices. When the ability of small farmers to provide for their households' consumption needs is compromised, they must make their living by working for others. The growth of the cash-cropping sector thus tends to be accompanied by an increasingly impoverished smallholder sector and a growth in wage labour. As adjustment leads to a decline in real wages, the food security of agricultural labourers is inevitably threatened.

Cuts in subsidies and the removal of price controls on consumer commodities such as food have caused particular hardship for urban dwellers, who have no land on which to grow their own food. In Sri Lanka, for example, following the removal of food subsidies, prices rose by 158 per cent for rice, 386 per cent for wheat flour and 331 per cent for bread between 1977 and 1984. In Zambia thousands of urban dwellers rioted when the price of their staple food, maize meal, rose by 50 per cent overnight in 1985. In addition to rises in consumer prices, incomes and employment opportunities have fallen steeply in the cities. Due to labour market deregulation and reductions in public-sector wages and employment, previously comfortable formal-sector workers have become the *nouveaux pauvres*. Many have turned to the informal sector to supplement their incomes. However, the fall in prices of informally traded goods and services, combined with a growing labour surplus, has eroded incomes in this sector. Many households have lost the purchasing power to meet minimum food needs. A study of low-income women in Dar es Salaam, for example, reported that 58 per cent of respondents had been forced to cut down from three meals a day to two. In Ghana in 1984 even upper-level civil servants could only afford two-thirds of the least-cost diet to meet nutritional needs (UNDP 1990).

The decline in household incomes in urban areas would appear to be directly related to the increase in women's participation in paid employment. As opportunities in the formal sector are limited, the vast majority of low-income women enter the informal economy as street traders, casual labourers, domestic workers or home-based workers. Most studies on the informal sector note that poor earnings are more often than not combined with excessively long hours. In rural areas, too, evidence suggests that an extra burden of work has been placed upon women within smallholder households through pressure to produce more crops for sale on the open market. Heavy workloads undermine the nutritional status of both women and their children. As hard physical work combined with repeated pregnancies increases women's nutritional requirements, where household income is too low to meet the family's consumption needs, and when cultural practices discriminate against women in the allocation of family food, their nutritional intake may fall below recommended levels. In addition to increasing women's susceptibility to infectious disease and to death and illness from pregnancy-related causes, this undermines the health of their children. Maternal malnutrition affects birth weight (a key determinant of a child's chances of survival) as well as the duration, quantity and quality of breastmilk. There is also concern that longer hours worked by women in adjusting countries will reduce the prevalence and duration of breast-feeding.

Adverse effects of adjustment on the health and nutrition of vulnerable groups have been recorded in several studies, though the worsening of social conditions is far from uniform. Some countries in South and South-east Asia, such as the Philippines, have been badly hit. However, this region has not experienced as marked a decline in living standards as Latin America and, most particularly, sub-Saharan Africa. For example, in 1990 the Commonwealth Secretariat reported that in seven out of nineteen Latin American countries for which data were available there was an increase in infant mortality rates (IMRs); in six countries the rates of improvement in IMRs had slowed; and in only six countries did IMRs continue to decline at the previous rate. It also found that seven countries in the region witnessed growing malnutrition among women and among children under 5 years of age.

Seven countries in sub-Saharan Africa are also reported to have experienced increasing rates of infant mortality between 1980 and 1985 (Kanji *et al.* 1991). However, the figures are difficult to interpret. UNICEF statistics indicate no increase in the rates of infant mortality, though the organization does accept that the rate of improvement has slowed down. Woodward (1992) also notes small decreases in infant mortality in sub-Saharan Africa.

He suggests, however, that if it were not for the reductions in infant deaths brought about by the introduction and use of oral rehydration therapy and the increase in immunization, IMRs in Africa would have been more or less stagnant in the 1980s.

Adjustment and urbanization

An important aim of economic adjustment policies has been to shift the economic focus from urban consumers to rural producers. It is assumed that, as this will counter *urban bias*, rates of rural–urban migration will be reduced. There are a number of problems with this assumption. First, as Chapter 7 makes clear, biases in the allocation of resources in LDCs work not so much against rural areas as against the poor. Thus, inequalities within cities are often far greater than those between urban and rural areas. A second problem with the IMF assumption is that rural dwellers have not benefited uniformly from adjustment policies. Rural wage labourers and small farmers unable to compete with the large cash-cropping sector have seen a fall in real incomes and their ability to survive. Many move to the towns and cities in search of alternative employment.

One consequence of the cuts in public expenditure demanded by adjustment packages is that there has been a reduction in the provision and maintenance of basic infrastructure. In low-income urban areas, where many Third World dwellers live in grossly substandard, overcrowded conditions, the lack of investment has resulted in the further deterioration in environmental conditions and a generally increased exposure to infectious disease. This, combined with the rapidly falling purchasing power of urban dwellers, makes it difficult to break the cycle of malnutrition and infection.

Although the plight of the urban poor has worsened under debt and economic adjustment, there is little evidence to suggest that rates of rural–urban migration have slowed. Indeed, the contraction of cash incomes in rural areas as a result of landlessness and inflation has encouraged urban drift in Africa (Sanders and Sambo 1991). Rural migrants account for a relatively small proportion of total urban growth in LDCs. However, migration can result in a new set of health problems for poor families and their communities. Large-scale migration, for example, may increase the labour supply in an area, thus depressing wages. Population movements can also lead to the introduction of new pathogens into urban environments as well as in the opposite direction. At the individual household level, migration is often associated with the disruption of family life. In sub-Saharan Africa, the migrant labour system taking men to urban, plantation and mining centres and separating them from their wives and families has been in existence since colonial times. What is new is the rapid rise in female migrants to African cities.

Female farmers in Africa produce the bulk of subsistence food, surpluses

58

in which are sold in local markets. They are, however, less involved than men in the production of crops for export. As the subsistence sector has experienced significantly lower price rises than export crops, women have hardly benefited from adjustment policies. Lacking access to credit and essential inputs, women have been vulnerable to take-overs by cash croppers. Those who arrive in urban areas in search of new employment find few opportunities for survival other than commercial sex. As described above, urban employment in the formal sector of adjusting countries has contracted significantly, and in the informal sector, where labour surpluses are threatening livelihoods, it is often difficult for women to obtain profitable work. As the economic attractions of prostitution have increased, so too have the risks of contracting and transmitting HIV infection. Several workers thus argue that the current AIDS epidemic in Africa has been aggravated by economic crisis and adjustment (Sanders and Sambo 1991).

Adjustment and health care expenditure

One of the most controversial issues in structural adjustment is its impact on health care expenditure. While the World Bank and IMF suggest that the health and education sectors are generally better protected than other public sectors during periods of expenditure reduction, evidence suggests that declines in health spending are significantly related to both degree of indebtedness and participation in adjustment programmes (Woodward 1992). Some countries have seen very substantial falls. In Zambia, for example, real expenditure on health fell by 22 per cent from 1982 to 1985, and the real value of the drugs budget in 1986 was only a quarter of the 1983 level (Stewart 1992). In Bolivia, central government health expenditure per person in 1984 was less than 30 per cent of that in 1980. The shortage of foreign exchange in many countries has reduced their ability to import medical supplies including basic drugs such as nivaquine and aspirin and consumables such as bandages, needles, syringes and sterilizing equipment. Due to the shortage of drugs in government facilities, patients issued with prescriptions have little choice but to buy their medicines from private pharmacies.

Cuts in health spending have had significantly negative effects on health indicators. In Zambia and Brazil there was a decline in child immunization programmes in the early to mid-1980s. In Ghana diseases that had been virtually eradicated by campaigns in the 1950s and 1960s, notably yaws and yellow fever, reappeared towards the end of the 1970s (Vickers 1991). Typhoid and hepatitis have increased in Chile, tuberculosis in Peru. The lack of health supplies combined with low wages has damaged the morale of health personnel, many of whom supplement their incomes in the private sector. As a result, standards of care in government health services in many adjusting countries have plummeted.

59

Cost recovery strategies are increasingly used by Third World governments to raise revenue to maintain their health services. However, evidence suggests that health service utilization declines with the introduction of user charges (Creese 1991). Under the cost recovery strategy adopted by the Ghanaian government, the nominal fee charged in 1983 for a first visit to a specialist clinic increased by a ratio of 400 by 1985. User fees were also introduced for casualty and polyclinics, drugs, laboratory tests, radiological examinations, medical and surgical treatment and medical examinations. Funds raised through fees allowed the government to improve its drug budget and other essential services. However, hospital utilization fell in many areas, as did child attendance at malnutrition clinics (Anyinam 1989).

The concept of cost recovery has been formally incorporated into international health policy through the adoption of the Bamako Initiative. Based upon the assumption that many people are willing to pay for treatment if they perceive a service to be of good quality, this UNICEF scheme provides loans for the purchase of essential drugs at cost. These are then sold at two or three times the cost price at the community level. The aim of the programme, which is implemented in African rural peripheries, is to create revolving funds which are capable of sustaining a continuous supply of essential drugs as well as providing additional subsidies for primary health care services. Once donor support is removed, recipient countries and communities are expected to have achieved self-sufficiency in drug purchases.

The Bamako Initiative emerged in response to the crisis in health care provision in Africa. Its supporters argue that it provides an effective way of strengthening community mobilization, rationalizing household expenditure on health, making more resources available for the improvement of service quality (particularly in mother-and-child health) and ensuring an adequate supply of essential drugs (thereby reducing irrational drug use). However, there is considerable debate about the ability of community health committees to manage sophisticated revolving funds. While UNICEF studies record very positive experiences of community financing and management (Jarrett and Ofosu-Amaah 1992), several commentators express doubts about the long-term viability of the scheme (Kanji 1989; Korte et al. 1992).

The most fundamental criticism made of the Bamako Initiative is that it appears to accept the shift towards the market-based allocation of health care. Efficiency is emphasized over equity, and in the poorest rural areas of a very poor continent accessibility to essential health resources has become increasingly dependent upon one's ability to pay. Concern about the economic ideology that underlies cost recovery relates to broader reservations about the shift during the 1980s from primary health care to a more selective approach (see Chapter 12). Advocates of selectivity deny that it represents a backward step in the philosophy of health and argue that, in times of resource scarcity, priorities in health interventions have to be made. It is widely recognized, however, that the promotion of child survival technologies is another factor that places extra demands on women's time.

Adjustment and education expenditure

A second sector to be adversely affected by cut-backs in public expenditure is education. In many adjusting countries, expenditure on schools has declined rapidly, as have enrolment rates. In sub-Saharan Africa, for example, per capita spending on education fell from $33 in 1980 to under $15 in 1986. Cut-backs in both recurrent and capital spending have resulted in poorly maintained school buildings, shortages of teachers, larger classrooms and a lack of textbooks, writing materials, blackboards and desks. In the face of increasing demands for parental contributions towards the costs of schooling, declining standards of education provide a disincentive for parents to send their children to school. In Nigeria, for example, school fees have been imposed on both primary and secondary education, leading to an immediate fall in enrolment rates. Even where education remains free, the costs of clothes, shoes, textbooks, bus fares and 'donations' to school fundraising efforts are beyond the reach of many poor households (UNICEF 1990). In Zambia, despite free primary education, parental expenditure in 1985 on basic items necessary for one child to attend primary school was more than one-fifth of average per capita income (Commonwealth Secretariat 1990).

At the same time that educational expenditure has declined in adjusting countries, real incomes have fallen and female participation in the workforce has risen. This places additional pressure on poor families to withdraw their children from school, either to work in paid employment or to help with household tasks including the care of younger children. The increase in child labour is a direct consequence of growing poverty in adjusting countries combined with the decline in educational opportunities. An estimated 80 million children between the ages of 10 and 14 undertake work which is either so long or so strenuous that if affects their normal development (UNICEF 1991). Underpaid and often underfed, many working children are exposed to insanitary conditions at work, toxic substances, accidents and sexual exploitation. An associated problem is the growing numbers of street children who, abandoned or unsupervised at home, are vulnerable to a range of health hazards including sexually transmitted diseases, drug abuse and crime.

The problems of child labour and child abandonment reflect the growing difficulties faced by women in managing their households. Large numbers of women have to join or to increase their participation in the paid labour force in order to make ends meet. However, cuts in government-supported social services combined with the rising costs of food and other consumption needs increase their domestic workloads. Many households respond to this growing pressure by withdrawing their daughters from school. As a result, female drop-out rates remain significantly higher than male. The long-term effects of a decline in female education are likely to be serious. As UNICEF (1983) suggests, it is usually a mother's level of education and access to information which will decide whether her children will be weighed and vaccinated; whether bouts of diarrhoea will be treated by administering food and fluids; whether weaning takes place at the right time; and so on. As the impact of maternal education on child health has been shown to work independently of household income, the current set-backs in female education will have repercussions on future trends in child survival.

ECONOMIC CRISIS: ALTERNATIVE STRATEGIES

The Bamako Initiative is part of a broader strategy promoted by UNICEF to protect the nutrition, health and education of vulnerable groups during the adjustment process. By the mid-1980s it was becoming increasingly clear that the health and welfare of the poor had significantly deteriorated. Thus, in 1987, UNICEF issued a two-volume study outlining its proposals for 'Adjustment with a human face'. While still accepting the need for adjustment, the 'human face' strategy called for a range of policies directed towards the poor. These included more expansionary macro-policies aimed at sustaining levels of production and employment; meso-policies designed to ensure a fairer share of incomes and resources; sectoral policies to support small-scale production, especially among small-scale farmers, the landless, urban informal workers and women; the restructuring of social expenditure towards basic needs provision; and special support programmes such as targeted food subsidies and public works employment schemes.

While the 'human face' policies relate to a number of the problems raised in this chapter (for example, the need to give greater support to small-scale farmers and the negative impact of adjustment on women), they do nothing to counter the international conditions that gave rise to economic crisis in the first place. Messkoub (1992), for example, suggests that 'adjustment with a human face' is an ineffective and partial palliative. Finding a contradiction between UNICEF's acceptance of charging health fees and its desire to aid the poor, he argues that small-scale, sporadic and 'targeted' efforts will do little to relieve poverty in the long run.

Unfortunately, there is no agreement about the solutions to Third World poverty. Proposals on the right of the political spectrum range from the current Western policy of limited debt relief and continued adjustment to 'debt for equity' swaps where debts are exchanged for shares in Third World industries. Left-wing theorists are more likely to propose that debt burdens are eliminated altogether and that fairer trade relationships are created.

Others evoke the populist model and suggest that LDCs 'delink' from the exploitative world market and follow more inward-looking development strategies. Kanji *et al.* (1991), for example, support policies which include an emphasis on regional self-sufficiency in food, expenditure towards agriculture and the social sectors and the development of indigenous technologies. It is clear that the solutions one finds to the economic crisis depend very much on political ideology and vested interests. The evidence presented in this chapter suggests that the objectives and impacts of adjustment policies have better served the West than the Third World nations they were meant to benefit.

CONCLUSION

The late 1970s saw the international health community adopt a highly optimistic approach to the health problems of less developed countries. The goal of 'Health for all by the year 2000' was widely endorsed, and primary health care was accepted as a radical new departure in health policy. Since this period, however, there has been an unprecedented decline in the health and living standards of the Third World poor. Economic crisis and growing debt made the 1980s a 'lost decade' for many poorer countries, particularly in Latin America and sub-Saharan Africa. This chapter has argued that the widespread implementation of economic adjustment programmes has exacerbated this decline.

The tragedy of economic adjustment is that the assumptions on which the model is based have long been contested. Decades of experience have demonstrated that economic growth does not in itself guarantee improvements in health and welfare. Indeed, as development theorists from Dudley Seers to members of the dependency school have pointed out, market-oriented policies benefit the rich *at the expense* of the poor and vulnerable. As the rich get richer and the poor get poorer in an increasingly unequal world, it is clear that the principles and goals of economic adjustment are far removed from those of health for all.

REFERENCES

Anyinam, C.A. (1989) 'The social costs of the International Monetary Fund's adjustment programs for poverty: the case of health care development in Ghana', *International Journal of Health Services* 19(3): 531–47.

Commonwealth Secretariat (1990) *Engendering Adjustment for the 1990s: Report of a Commonwealth Expert Group on Women and Structural Adjustment*, London: Commonwealth Secretariat.

Cornia, G., Jolly, R. and Stewart, F. (1987) *Adjustment with a Human Face*, Oxford: Clarendon Press.

Creese, A.L. (1991) 'User charges for health care: a review of recent experience', *Health Policy and Planning* 6(4): 309–19.

Hewitt, T. (1992) 'Developing countries: 1945 to 1990', in T. Allen and A. Thomas (eds) *Poverty and Development in the 1990s*, Oxford: Oxford University Press, pp. 221–37.

Jarrett, S.W. and Ofosu-Amaah, S. (1992) 'Strengthening health services for MCH in Africa: the first four years of the "Bamako Initiative"', *Health Policy and Planning* 7(2): 164–76.

Kanji, N. (1989) 'Charging for drugs in Africa: UNICEF's "Bamako Initiative"', *Health Policy and Planning* 4(2): 110–20.

Kanji, N., Kanji, N. and Manji, F. (1991) 'From development to sustained crisis: structural adjustment, equity and health', *Social Science and Medicine* 33(9): 985–93.

Korte, R., Richter, H., Merkle, F. and Gorgen, H. (1992) 'Financing health services in sub-Saharan Africa: options for decision makers during adjustment', *Social Science and Medicine* 34(1): 1–9.

Messkoub, M. (1992) 'Deprivation and structural adjustment', in M. Wuyts, M. Mackintosh and T. Hewitt (eds) *Development Policy and Public Action*, Oxford: Oxford University Press, pp. 175–98.

Sanders, D. and Sambo, A. (1991) 'AIDS in Africa: the implications of economic recession and structural adjustment', *Health Policy and Planning* 6(2): 157–65.

Seers, D. (1969) 'The meaning of development', in D. Lehmann (ed.) *Development Theory: Four Critical Studies*, London: Frank Cass.

Stewart, F. (1992) 'Can adjustment programmes incorporate the interests of women?', in H. Afshar and C. Dennis (eds) *Women and Adjustment Policies in the Third World*, London: Macmillan.

United Nations Children's Fund (1983) *The State of the World's Children 1984*, Oxford: Oxford University Press.

—— (1990) *The State of the World's Children 1990*, Oxford: Oxford University Press.

—— (1991) *The State of the World's Children 1991*, Oxford: Oxford University Press.

United Nations Development Programme (1990) *Human Development Report 1990*, Oxford: Oxford University Press.

Vickers, J. (1991) *Women and the World Economic Crisis*, London: Zed Publications.

Woodward, D. (1992) *Debt, Adjustment and Poverty in Developing Countries*, 2 vols, London: Frances Pinter in association with Save the Children.

World Bank (1990a) *World Development Report 1990*, New York: Oxford University Press.

—— (1990b) *World Development Tables 1990–91*, Washington, DC: World Bank.

4

TRADITIONAL MEDICINE
Its extent and potential for incorporation into
modern national health systems

B. Hyma and A. Ramesh

INTRODUCTION

On the basis of a resolution adopted at the 1977 World Health Organization
Health Assembly, many developing countries have taken action to develop
policies and programmes for the integration of traditional systems of
medicine into national and primary health care systems. In 1978 the
International Conference on Primary Health Care (PHC), held in Alma Ata
under the sponsorship of the WHO and UNICEF (the United Nations
Children's Fund), passed additional resolutions to implement its strategy of
attaining 'Health for all by the year 2000'; these resolutions promote the
incorporation of useful elements of traditional medicine (TM), as well as its
practitioners, into national health systems. It is well documented that
pluralistic medical services exist in many parts of Africa, Asia and Latin
America (Leslie 1976; Bannerman *et al.* 1983; Leslie and Young 1992). The
recent shift among international donors and in national policy in favour of
providing support for TM and incorporating it into the dominant modern/
Western health care programmes stems from a range of factors. For example,
the WHO observes that the last decade has seen considerable growth
in popular, official and commercial interest in the use of traditional
practitioners and their remedies. Indeed, for the majority of the world's
population these have been, and in many instances are still, the only forms
of medical care readily available. In the 1980s, indigenous health care
resources (traditional practices, remedies and practitioners) were thus
increasingly brought under the purview of more formal health services.

There are other practical and philosophical reasons for the increasing
popularity of TM. The economic crises and political disruptions affecting
many developing countries have often led to severe shortages of certain
modern drugs which are often imported, thereby forcing more people to
make use of TM whether or not this is government policy. In the worst of
situations, this could lead to increased quackery and charlatanry, especially
where TM is intimately linked with the essential elements of PHC (WHO

1985). Virtually all governments strive to bring modern/cosmopolitan medicine and health care to their populations. In developing countries, the means and resources to do this often do not exist. The recurrent and capital costs of expanding modern health care based on (exogenous) Western models of development to reach rural populations are enormous and exploding (Phillips 1990). Comprehensive primary health care provides a strong foundation for full use of the existing material and human resources of the community, thereby hopefully setting an alternative trend towards the mobilizing and revival, or at least strengthening, of indigenous culture, knowledge, practices, resources and technologies, as well as of the organizational forms to sustain local and regional development.

Against such a background, this chapter examines the process of the integration of TM into modern national health systems from both theoretical and practical points of view. The WHO's recent review of national experiences in the 1980s indicates that there is no single universally accepted concept of integration, and many countries have yet to clarify it in their public policies. The second stage of this discussion therefore provides a brief overview of policies, strategies, achievements and progress, with particular focus on the experiences of selected Asian countries in the 1980s. Particular advantages and fundamental problems of integration are explored. The extent to which governments have gone to promote institutional integration in recent years leads to our third objective: to examine appropriate policies and strategies that have been identified as useful in promoting integration at regional, national and community levels and among professionals. Key issues and ways and means of achieving full integration are explored.

Finally, we search for a new medical paradigm that will reflect an ideal type for the integration of modern medicine (MM) and traditional medicine (TM). What is the right kind of relationship in an integrated model – one where TM plays a complementary, supplementary and subordinate role, or a co-operative and mutually supportive role? Furthermore, the 'integrationist', 'purist' and 'structured coexistence' debate also needs to be re-examined. This chapter suggests that to date the process of unification or merger of the two systems still faces many ambiguities.

THE UTILIZATION OF TM AND TM PRACTITIONERS IN NATIONAL HEALTH PROGRAMMES TODAY

In many Asian countries, health care systems have been highly diversified and pluralistic for over a century. Several forms of empirical indigenous medicines are found (Leslie 1976; Good 1987; Phillips 1990; Leslie and Young 1992). Medical pluralism can denote the coexistence of multiple systems of medicine (traditional, modern, folk) giving multiple choices to individuals, or it can even mean 'pluralism within a particular system, allowing access to various levels and types of care' (Phillips 1990: 75). In most developing countries, the presence of dualistic medical systems has resulted from specific historical events.

The introduction of Western medicine during the last century or so has resulted in a coexistent or complementary situation. While traditional medicine flourished mostly in the private sector, the nineteenth-century introduction of Western science and Western-oriented medical care has come to dominate and control institutions, supplemented by other health care systems. The nature of indigenous medical health care systems as they evolved over time is highly differentiated with regard to principles of treatment, diagnosis and practices. It is influenced by differences in culture, religion, levels of social and economic development and other characteristics. In India, Pakistan, Indonesia and South Korea, for example, TM is already well established in the private sector, and certain institutions have gained both national and international reputations for the manufacture of herbal remedies. The policies of many modernizing states have also given rise to different levels of performance in traditional medical systems. Medical pluralism is now generally viewed as a positive factor in the health care programmes of many developing countries (Akerele 1987; Good 1987).

The experience of WHO-participating countries

The professional and political acceptability of TM varies from country to country and even within individual countries. It might be expected that the integration of TM and its practitioners (TMPs) would be supported by government legislative policies. Table 4.1 provides examples of the mid-1980s status of TM in selected Asian countries, based on criteria used by the WHO to investigate the process of integration of TM into primary health care. The WHO identifies four broad categories of existing policies and legal regulations on TM, which are not necessarily rigid or conclusive and have some overlap:

1 *Complete prohibition by restrictive legislation*: only the practice of modern, scientific medicine is recognized as lawful; enforcement varies from country to country.
2 *Toleration and non-intervention* (e.g. in Germany, United Kingdom, some Latin American countries, Thailand, Indonesia, Hong Kong, Egypt, Sudan).
3 *Formal recognition* (state-regulated) following training, registration and licensing (Swaziland, Zimbabwe, Thailand, India, etc.).
4 *Formal recognition and official sanction* to integrate TM in the formal health system: integrated training of practitioners is official policy (e.g. in China, Nepal, Vietnam, PR Korea, Sri Lanka, India, Pakistan).

In South Asia, traditional systems of medicine are often formally institutionalized, with extensive facilities for education, training, research and

Table 4.1 Integration of traditional medicine and primary health care, selected Asian countries

	China	Korea	Vietnam	Philippines	Fiji	Solomon Islands	Papua New Guinea	Malaysia	Singapore
1 Population*	1,004,470	38,197	53,257	48,511	625	231	3,164	13,660	2,403
2 Policy	+	+	+	o	o	+	o	o	o
3 TM practice	+	+	+	+	+	+	+	+	+
4 Quantification	+	+	+	o	o	o	+	o	o
5 Formal structure	+	+	+	o	o	o	o	(e)	(g)
6 TM education with MM	+	+	(a)	o	o	o	o	(e)	+
7 TM integrated with MM	+	o	G	o	o	o	o	(e)	o
8 Sponsored training	G	G	G	G & P	o	o	o	(e)	P
9 Legal attitude	S	N	S	N	N	S	N	N	N
10 Professional attitude	S	N	S	N	N	S	N	N	A
11 Economic need	+	o	+	+	+	+	+	+	o
12 Cultural need	+	+	+	o	+	+	+	+	+
13 Geographical need	+	o	+	+	+	+	+	+	o
14 Formal research	+	+	+	+	o	o	+	(f)	+
14a Research validation	+	+	+	(b)	o	o	(c)	o	+
14b New knowledge	+	+	+	(b)	o	o	(c)	(f)	+
14c Local identification	+	+	+	(b)	(b)	o	o	o	o

Source: Based on WHO (1984)

Key:
+ Positive; yes; characteristics present; procedure and/or programme exists.
o Negative or absent; no.
G Government.
P Private.
S Supportive.
N Neutral.
A Antagonistic.
* Population, 1980, estimated (000s).

(a) Limited to traditional birth attendants (TBAs).
(b) Herbal remedies only.
(c) No transfer of information; local use only.
(d) Botanical and descriptive studies.
(e) TBAs only.
(f) Descriptive studies.
(g) Chinese medicine only for ethnic population segment.

health care delivery, including hospitals, clinics and pharmacies exclusively designed for these branches of traditional medicine (Bannerman *et al.* 1983; WHO 1985). In India, for example, the government provides considerable financial and institutional support for the development of Ayurveda, Unani and Siddha systems of medicine. With the enactment of the Indian Council Act of 1970, the Indian government declared its intention to use the indigenous system in its national health services (Ramesh and Hyma 1981). The Central Council of Indian Medicine was established in 1971 by Act of Parliament to evolve uniform standards in education, to make mandatory the maintenance of a central register of Indian medicine and to promote research in traditional medicine (WHO 1985). Sri Lanka established a separate ministry of TM; pilot projects have been undertaken to involve certain traditional practitioners in local PHC programmes. In Nepal, so far, PHC networks for the modern and Ayurvedic systems remain separate and not integrated.

Traditional medical practices in Japan, South Korea, Fiji, the Philippines, the Solomon Islands and Papua New Guinea are so far independent of government health services. In Africa, in the 1980s, many experimental projects were working towards establishing linkages between TM and MM. Several studies have been reported from Ghana, Nigeria, Swaziland, Zaire and Zimbabwe. Zimbabwe has recently made considerable progress in ensuring collaboration between traditional health practitioners and the official health system – for example, by the organization of joint seminars, the publication of a register of traditional practitioners and some forms of self-regulation (WHO 1985).

The Chinese experience in combining two vastly different systems of medicine to obtain comprehensive health coverage for its population is widely admired. The WHO observes of Chinese TM that it is a formal, structured system. It has a long history and tradition, with its own colleges, research institutes and disciplinary controls. Modern technology and science are applied to the traditional system, and the Chinese government intends to develop legislation to promote further collaboration at all levels of the health system and in all its functions: in health care delivery, research, education and training, and in standardization and quality control of traditional remedies. There is also strong financial support for these activities. However, it has taken concerted action to achieve integration, and attempts to integrate the two systems of medicine failed in the period since 1949. Traditional Chinese medicine therefore had to develop as an independent and parallel system of medicine before a stage was reached for mutual respect. Today, every Western-style medical school in China contains a department of traditional medicine, and every school of traditional medicine contains a department of modern medicine; practitioners of both modern and traditional medicine are employed in modern hospitals, and they also work together in the commune health centres. The policy of the Chinese

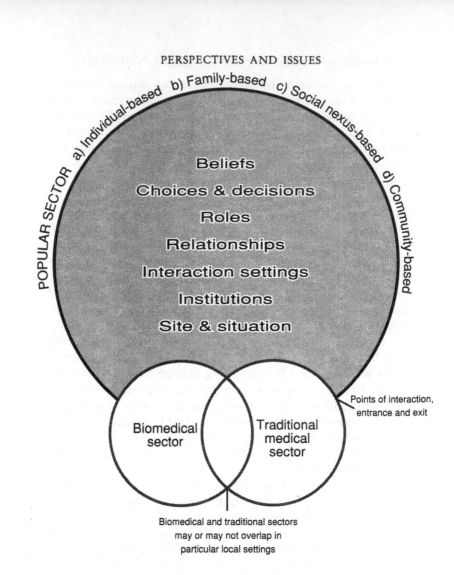

Figure 4.1 An ethno-medical system
Source: Based on Good (1987: 24)

health system has always been to use those elements of ancient Chinese healing that are effective and to discard those which are not (WHO 1985).

Several Asian and African countries now consider the concept of integration a reality that could be fully achieved in the near future. Phillips (1990) thus argues that the acknowledgement of medical pluralism by researchers, biomedical professions and governments means that modern/Western medicine need not automatically or uncritically be regarded as the norm, and local beliefs need not be seen as an impediment to improved health care.

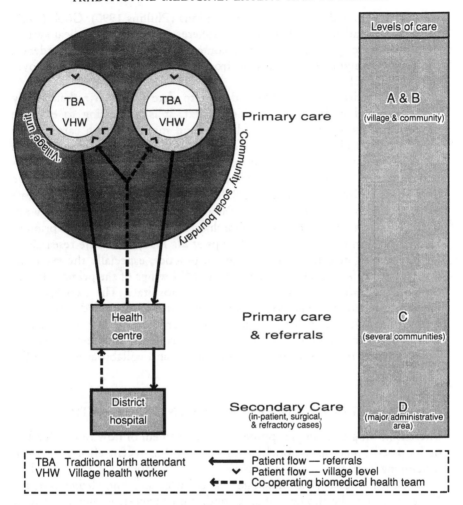

Figure 4.2 Structure of proposed co-operative health care system
Source: Based on Good (1987: 313)

THE INTEGRATION DEBATE

An ethno-medical system may be defined as comprising the acquired knowledge, resources (formal and informal), organization, behaviour and strategies (traditional and scientific, indigenous and imported) that a community or society utilizes for individual or collective well-being (Good 1987). An ethno-medical system thus may be seen as evolving from a popular view of the world and life which incorporates theories and accepted explanations of the nature and causes of illness and the appropriate and expected remedies. It is therefore very much influenced by social and

71

cultural factors and by interpersonal relations (Phillips 1990). Good (1987) presents a broad model of three main spheres of an ethno-medical system with different degrees of interaction, contact, overlap and interdependence, the relative weight of each varying from culture to culture and also over time (Figure 4.1).

The interaction between TM and MM to date has been rather poorly researched and understood. In this context, it is interesting to examine Good's models for integrating traditional and modern health care (Figure 4.2). Good's model presents a type of integration that might be appropriate for agrarian communities in countries such as Kenya, with their community-based preventive and curative care. In south India, official attempts have been made to integrate traditional Siddha medicine with MM in the same primary health centre complex in Tamil Nadu, although there are weak linkages and referrals to the next hierarchical level. The Tamil Nadu primary health centre model is operative (Phillips *et al.* 1992); but little research so far has investigated the referral system in practice, especially the extent of linkages and referrals between the TM and MM wings of the primary health centres, and *taluk* (subdistrict) and district hospitals. The two types of medicine are largely separate, although working in a unified institution. Unlike the Chinese model of integration, problems relate to different techniques of the two systems in diagnosis, referrals and treatment. Research is also needed into the attitudes, co-operation and collaboration of TMPs and MMPs.

CONCEPTS AND PRACTICES IN INTEGRATION

There is no single or simple approach to the problem of how to involve TM or its practitioners in national health systems, but recent efforts by some developing countries to effect institutional integration are recognized as important in the survival of indigenous cultures and knowledge systems. Integration may be thought of as a continuum between minimal and total integration, since nowhere does there exist totally successful and complete integration of the different medical systems. Indeed the term 'integration' can be defined in different ways, to mean, for example, institutional integration through national health services; consumer use of more than one medical system; and health workers providing a combination of traditional and modern care. A fourth form of integration relates to the cognitive integration of TMPs and MMPs on an interpersonal basis.

Integration through national health services

The institutional integration of TM is a process in which formal health services begin to incorporate and work with TM for the benefit of the whole population. TM is no longer ignored by the official sector. The WHO identified factors that might promote the integration of TM in PHC when supported by legislative policy. The scarcity of resources (personnel, buildings, materials and finances) in PHC services is a factor that encourages integration and can improve its prospects. Generally, full integration occurs through primary medical resources (drugs, techniques and empirical knowledge in PHC). With the exception of the Chinese experience, there are as yet no real examples of full integration of medical personnel. What is generally found is forms of structured coexistence or co-operation (within mutual referral systems) between cosmopolitan and traditional medicine. For example, integration of traditional health personnel seems possible in obstetrics without very many complications. Certain case-studies in Africa indicate a possible division of responsibilities between Western-trained doctors and traditional healers (for example, in treating fractures or poisoning and in psychotherapy).

Cai (1988: 527) reports that 'Chinese medicine in both its theoretical and practical aspects is now undergoing a process of applying modern scientific methods, and is developing a new integration with western medicine.' Researchers working on Chinese medicine are trained in subjects such as physiology, anatomy, biochemistry, pathology, physical diagnosis, laboratory diagnosis, microbiology, immunology, radiology and molecular biology, as well as in the use of modern scientific techniques such as electromicroscopy and chromatography. The clinical and scientific attainments of integrated medicine have convincingly demonstrated that this is an appropriate orientation for the development of China's medicine. This does not mean that the traditional medicine has suddenly lost its vitality; it has always had a high capacity for adaptation and the absorption of new stimuli (Cai 1988).

Advantages of institutional integration

The WHO has identified several advantages of the integration of TM and MM systems. Integration can offer reciprocal benefits to each system; it can improve the general health care knowledge for the greater human welfare, especially in view of the inherent possibilities for wider and more efficient population coverage; it can enhance the quality of the practitioners, as well as increasing their numbers; and it can promote the dissemination of knowledge relating to primary health care. Above all, it probably offers the best means of achieving the goals of health for the entire population by the year 2000.

Drawing on field experiences and observations of the integration process in selected primary health centres in Tamil Nadu, south India, we offer specific observations on the institutional integration of MM with TM. Legislation and policies are being introduced to promote the integration of traditional medicine in primary health care. In principle, this protects patients from substandard care and the community from charlatans. Government support is being given for the programme; a formal organizational structure (a national comprehensive strategy) has been established and a regulatory body created. The scarcity of resources is a factor which encourages institutional integration in a common set of facilities.

Other indirect benefits of integration stem from the development of centres of modern research, training and education in indigenous medical systems. These centres provide training for personnel, undertake clinical and pharmacological research and scientifically evaluate and validate existing traditional medical practices and remedies. They can also help to improve therapeutic efficiency, standardization, quality control, distribution and availability of herbal medicine. Integration can also promote protection, conservation and propagation of medicinal/herbal resources.

A Siddha science development committee established by the Tamil Nadu government in 1981 was charged with a range of tasks, including: recommending regional pharmacies for the production and supply of medicine; testing of standard raw drugs purchased by the pharmacies; supervising the copying of rare manuscripts (often from ancient palm-leaves and texts) and the reprinting of rare books; translation of eminent Siddha literature into English; systematizing the materia medica; and publishing a Siddha encyclopaedia. This work is of a long-term nature, and low priority is given in terms of resources needed to accomplish the tasks. Yet TM is changing and growing within a state-established infrastructure. It is to be hoped that modern and traditional practitioners will meet, co-operate and collaborate to define their range of activities, to develop referrals and to learn from each other's approaches, experiences, knowledge and practices, as has happened in China. Integration in India would offer patients the fullest possible range and choice of medical treatment. Finally, institutional integration of TM could promote the revival of Tamil culture, heritage, regional identity and development, since medical integration implies an enquiry into the proper place and conditions of indigenous knowledge and resources in a society.

Criticisms of institutional integration

Empirical evidence indicates that TM is still being given a low priority in the quality of national health care programmes and that it almost everywhere occupies a subordinate position in the provision of public health care facilities (with the exception of China). Conceptually, integration is often seen as a process of *modernizing* TM, thereby implicitly validating the supposed superiority of Western medicine. TM seems to occupy an inferior, subordinate role in most of the primary health centres studied in India, and it is beginning to lose its autonomy. Even though integration implies a harmonious coexistence of both or all systems of medicine, in reality this is still very ambiguous.

It may be argued that integration does not necessarily mean a complete blend of TM and MM. Indeed, traditional medical development in India (and even in China) does not assume a real synthesis of indigenous and modern medicine; interaction takes place only at the level of practice. For example, while in China the traditional technique of acupuncture is used for anaesthetic purposes, its theory does not integrate with modern medical concepts. In India, each system develops separately. However, modernization of TM in India under government protection policies has taken place with similar results to those in modern medicine, and emerging structural problems are also similar. TM is forced to assume the qualities of MM in the integration of public medical and health services and is facing the same structural problems as MM in the transition process, rather than retaining its own individual, rich cultural heritage and its body of theoretical knowledge and practices (Phillips *et al.* 1992).

The integration of traditional healers into national health systems almost always means the incorporation of traditional medical knowledge by grouping and supervising its practitioners into special categories of 'auxiliaries', and not all are prepared to accept this. Indeed, the social status accorded to a traditional healer within a given culture is frequently likely to be higher than what can be achieved through recognition by the public administration and by integration into the national health system as an 'auxiliary'.

In India, large-scale systems are dominated by the medical profession, which is an integral part of an urban, modernized middle class. In spite of state policy supporting TM, the indigenous tradition continues to be relegated to an inferior paramedical, para-professional status. However, professionalized TM (such as Ayurveda and Unani) is based on emulating the Western model and has developed its own network of research, institutions, professional associations and publications. Pre-professionalized Ayurveda (mostly village practitioners) does still exist but is being eroded by professionalized TM. Indigenous systems of medicine remain divided or unevenly served because of class, caste and religious and ethnic social segments, which perhaps argues against generalized social acceptability (Neumann and Lauro 1982).

The cognitive integration of modern and traditional personnel

It might be argued that an individual cognitive capacity is the most important determinant for effective integration; that is, awareness both of techniques and of philosophy/theory behind the cause of illness is necessary for effective prevention and curative treatment. How well integration occurs will be measured by the extent to which TMPs and MMPs acquire and exchange skills and knowledge (theoretical and practical), and work and share without imposing on one another and without submerging individual characteristics. It should be emphasized that the two medical systems are based on different philosophies, theories, histories and geographies, different aetiologies of disease, educational and training backgrounds, and diagnostic and treatment methods. As a result, designing and achieving an ideal form of integration may not be simple or even possible. To be effective, integration depends on the revival, availability, accessibility and development of traditional knowledge, as well as on the simplification of modern medical knowledge. In this context it can be shown that synthesis (via the merging of TM and MM) may not yield the optimum benefit to health and health care.

Ideally, it is often hoped that traditional medical knowledge, although it can be used in modern practice, should not simply be co-opted into the system of modern knowledge. Being separate from MM will allow TM to retain its own identity, which in the long run, given sufficient scientific research, political will, commitment and planning, may lead to the realization of a new paradigm for health care. The merging of knowledge might only lead to the dominant one (MM) taking over and eventually distorting the natural features and identity of TM.

Thus the degree of complementary relationships may be seen in the cognitive integration of knowledge in individual practitioners. Yet in practice, not every aspect can be complementary. Problems of working together and the extent and benefits of referral systems are not yet clearly researched or understood. The ability to integrate knowledge arguably rests more heavily on modern health personnel, who must first develop an understanding of traditional knowledge. The current practice tends to be to train TMPs in background knowledge of modern medicine in order for them to be able to make appropriate decisions about cases that need to be referred to hospitals. Such decisions are made on the basis of limitations in providing treatment. The limited confidence in the improvement of traditional knowledge can suppress the full potential role of TM within the framework of PHC. Consequently, the prevailing trend is for TMPs to play a supplementary role in health care systems.

Major questions remain. Is it possible to establish an *ideal* form of integration? What is the right kind of relationship between TM and MM? The search for a new medical paradigm reflecting an ideal type of integration persists in the process of health care planning; the dilemma, debate and discussion continue to occur without any clear-cut initiative towards resolution.

A review of the empirical process of integration in India and China reveals a number of features. In India, formally qualified and trained TMPs in government institutions are used in PHC services. These TMPs have a background knowledge of MM. By contrast, doctors trained in Western medicine are rarely given background knowledge or training in TM. Thus to date, the integration of TM's (theoretical) knowledge into government programmes in MM institutions has not been attempted. Having TMPs work in established primary health centres and in district and *taluk* hospitals appears to imply a subordinate, supplementary and probably inferior role. Differences in educational background and fields of interest, and different licensing requirements between TMPs and MMPs, imply limited integration. Practitioners seem somewhat limited legally and confined to their own specific areas (Phillips *et al.* 1992). Furthermore, India's government so far has been unable to develop suitable procedures or training programmes to utilize or recruit existing traditionally trained or hereditary practitioners into formal health systems.

In China, by contrast,

> the decision to combine Chinese medicine and modern western medicine at all levels of health services in medical research has led to the evolution of a new form of 'integrated medicine' which seeks to apply the best of both systems, and to offset the weaknesses of each. In this way, the two systems are being gradually fused together. But it is a huge task which will take a long time.
>
> (WHO 1985: 34)

The process of 'integrated medicine' in China is said to involve 5,000 doctors who have learned traditional medicine for more than two years. Increasing numbers of experts and scholars of Western medicine have begun to appreciate and to study traditional medicine, actively taking part in the research on the combination of the two medicines. These doctors and experts apply the strengths of the two medicines in disease prevention, medical treatment and medical research.

The WHO's attitude is that integration calls for the use of modern knowledge and techniques in the application of TM. This implies a number of things. Efforts should be made to inherit, explore and organize TM and to raise its levels. Doctors of TM should be united and relied upon, and doctors of Western medicine should be organized to learn and study TM; a combination of the two medicines should be promoted. Advanced technology should be introduced and research in TM modernized. The course of TM and the combination of TM and Western medicine should develop in tandem and the necessary equipment be provided for their development. Finally, medicinal resources should be protected and well utilized (WHO 1984).

Integrated medicine and professionalization

It can be said that professionalization in India and China has had different effects on health care development in each. In India, a revivalist movement of indigenous/traditional medical systems stemmed from two ideologies, integrationist and purist (Leslie 1976), albeit with the same aims, that is, the professionalization of integrated medicine. Both ideologies were led by the urban middle class. The integrationist school set out to supplement TM with modern medical knowledge, leading to its virtual submersion under the structure of modern medical care systems. This school made full use of modern diagnostic facilities, pathology and surgery (including bacteriology and virology). The purist school, which opposed the new breed of formally trained TMPs, intended to preserve and nurture the old institutions; it concerned itself with interpreting, verifying and validating the huge body of theoretical and practical knowledge in the materia medica. The new breed of TMPs are creating new social organizations which may become a new profession in health services. A newly established college will stress and preserve primarily the knowledge of pharmacology and herbology and undertake clinical research, putting decreasing emphasis on indigenous medical systems. However, in South Asian and South-east Asian countries (and more recently in Africa), there are numerous special institutions and associations for traditional forms of care in the private sector; these can continue to work effectively and autonomously without becoming part of an integrated health service. Their roles and characteristics still remain to be widely researched and understood.

Integration at the level of the individual consumer and health practitioner

Integration can occur spontaneously through consumer choice of services in a pluralistic medical system. Patients can move from one system to another or use several systems simultaneously. Integration occurs with individual preferences in medical care based mainly on a perception of the effectiveness of treatment and the nature of illness and disease. People may also seek health care services from different sources because of costs, time needed to obtain care, expectations of long-term cure, previous knowledge, beliefs, familiarity, experience and social contacts. In modernizing societies, consumers have very often combined different types of health care services appropriate to their situations. Government intervention can therefore be seen as focusing on the creation of an atmosphere in which each type of medical care is developed to the highest possible quality (Akerele 1987; Phillips *et al.* 1992).

Integration through health workers

Very little is currently known about the role of community health workers in providing linkages between traditional medicine and biomedical establishments. Primary health care workers rarely seem to possess knowledge of herbal medicine, nor are they trained in its utilization. Without sufficient knowledge and awareness of home remedies, common medicinal plants and their therapeutic applications, health workers may find it difficult to gain the confidence of the community. They will also lack clear insight into what community members do for themselves in times of sickness.

A major aspect of TM lies in its holistic and ideological philosophies. In China, for example, professionalization did not receive any more attention than did the utilization of indigenous knowledge in the health care systems. To some extent the erstwhile 'barefoot doctor' concept demonstrated the success of the adoptive efficacy of indigenous as well as modern medicine.

Good (1987) suggests that popular knowledge of health care should still be the first priority, because it is neither ethical nor wise to impose a single choice of health care on people. The revival and support of TM institutions are, in one sense, allied to the production of alternative knowledge of health care for sustainable development. Furthermore, this stresses a very important aspect of the indigenization of development processes. It can be argued that the promotion, production, improvement and preservation of indigenous health care knowledge and practices involve commitment and responsibility not only from traditional institutions but also from the scientific community, medical professionals and planners. This in turn not only requires that traditional indifference and mistrust are overcome, but also that recognition, mutual trust, flexibility, respect and collaboration among the practitioners of the various systems concerned are ensured.

POLICIES, STRATEGIES AND PROGRAMME GUIDELINES FOR INTEGRATION

Relatively few countries have taken serious steps to integrate TM and MM nationally on a large scale. The WHO has identified guiding principles for such an endeavour and stresses that government commitment to TM as a realistic option is an essential ingredient for success. It suggests a long-term plan of action to cover the following areas: development of a government policy; introduction of enabling legislation; funding of pilot projects; budgeting for training programmes; and providing support at institutions of higher learning for research into the theory and practice of integration at the primary health care level.

The formulation of appropriate policies and strategies to promote integration is recommended at various levels: international/regional (with WHO collaboration and support); national (e.g. technical and financial support, legislation, formal training, registration, medicinal plants, etc.); professional (e.g. institutionalizing practices, professionalization of TM); and community levels (training of village or community health workers). The WHO's recommendations to achieve full integration cover a wide range of activities, including: legalization by government of the integration of TM into the present health care system at institutional and community levels; research and full documentation of traditional medicine and medicinal plants; registration of traditional medical practitioners; establishment of an institute of traditional medicine; training of traditional medical practitioners and traditional birth attendants; cultivation of medicinal plants; integration of the two systems of medicine at all levels of the health service; the granting of equal status to practitioners of the two systems; and the training of village or community primary health care workers in both traditional and modern medicine.

CONCLUSION

The institutional integration of TM into existing national health services as it is being implemented by governments of developing countries presents certain challenges and unresolved questions. The results of incorporating TM have so far yet to be evaluated or examined comprehensively. Full integration does not seem to be occurring at the present time. Integration seems to be proceeding only on a limited scale, and the important issue of replicability on a large scale is unknown. In general, the WHO identifies four major areas needing improved performance in the primary health care integration process: management; training; research; and information generation, storage and dissemination. All these needs are evident in a study by Phillips *et al.* (1992), who explore a preliminary process for an integration model at the primary health centre management level. Managerial aspects of the primary health care system need particular strengthening through expert committees and advisory group meetings at regular intervals. Since evaluation should be an integral part of all projects, there is a need to develop suitable monitoring and evaluation procedure at various levels.

Research is vital for more effective implementation of TM in primary health care. Even though scientific (validation) research is currently being carried out in many Asian countries, there has been very little useful applied research so far into the theories, practices and pharmacology of TM. Research that has been conducted has mainly been at the central and state levels in metropolitan areas, in isolation from the rural and lower levels of the health care system.

Basic studies should include at least three major foci:

1 Research on a small scale in selected primary health centres, in small administrative units where both MM and TM personnel function. Pilot projects can demonstrate process and feasibility and provide data for evaluation; successful projects can then be identified, replicated and perhaps incorporated in other primary health centres.
2 Comparative studies involving TM, MM and combined MM and TM modes of therapy and the forecasting of undesirable effects.
3 Qualitative research and investigations, such as epidemiological studies of community behaviour and the impact of integration on the surrounding communities. These could become a part of wider surveys relevant to community development. Knowledge, attitude and practice studies of the surrounding community or special groups (such as local healers) are particularly useful in integration projects.

Education, training and personnel development are also essential to hasten, improve or expand the integration of TM in the primary care setting. To date, only very limited programmes (such as workshops, consultations, exchanges, conferences, study tours, short courses and the distribution of educational materials) are available to administrators, to practitioners of both forms of medicine and to other health workers and staff of primary health centres involved in the integration process. Encouragement for the inclusion of TM in general education at all levels and in modern medical curricula, as well as the recording and disseminating of actual integration experiences, would also stimulate interest.

No system of medicine is perfect, and every system has its own contribution to make. Arguably, official policies should help to develop and to utilize the best of all systems of medicine for providing effective and efficient medical care for the population. It can be considered a liberalized approach in principle to provide a *choice* of treatment based on individual preference. The contemporary growth of indigenous medical institutions may actually strengthen the continued existence of such a liberalized system.

However, the *quality of care* provided by TM has not yet been fully assessed. Low-priority political and financial commitment and the lack of innovative design, planning, evaluation, monitoring, management and training all have to be overcome. The assessment and maintenance of good-quality of care are essential if TM and MM are to achieve integration; one cannot be the poor relation in quality terms.

The practical value of indigenous medical systems and their persistence have already been well demonstrated by many studies. Whether one should support the development of a dual or autonomous health care system or an *integrated* system appears an endless debate in the integration process. In a dual system, consumers can select modern or traditional services; while in an integrated system, TM may continue to occupy a subordinate role to 'superior' MM and lose its unique/individual cultural characteristics. Overall, the process of unification or merger of the two systems of modern and traditional medicine still faces many ambiguities, although new attitudes at all levels may in the future enhance the process of integration.

REFERENCES

Akerele, O. (1987) 'The best of both worlds: bringing traditional medicine up to date', *Social Science and Medicine* 24(2): 177–81.

Bannerman, R.H., Burton, J. and Ch'en, W.C. (1983) *Traditional Medicine and Health Care Coverage*, Geneva: WHO.

Cai, Jingfeng (1988) 'Integration of traditional Chinese medicine with Western medicine – right or wrong?', *Social Science and Medicine* 27(5): 521–9.

Good, C.M. (1987) *Ethnomedical Systems in Africa: Patterns of Traditional Medicine in Rural and Urban Kenya*, New York: Guilford Press.

Leslie, C. (ed.) (1976) *Asian Medical System: A Comparative Study*, Berkeley, CA: University of California Press.

Leslie, C. and Young, A. (eds) (1992) *Paths to Asian Medical Knowledge*, Berkeley, CA: University of California Press.

Neumann, A.K. and Lauro, P. (1982) 'Policy and evaluation perspectives on traditional health practitioners in national health care systems', *Social Science and Medicine* 16: 1825–34.

Phillips, D.R. (1990) *Health and Health Care in the Third World*, London: Longman.

Phillips, D.R., Hyma, B. and Ramesh, A. (1992) 'A comparison of the use of traditional and modern medicine in primary health centres in Tamil Nadu', *GeoJournal* 26(1): 21–30.

Ramesh, A. and Hyma, B. (1981) 'Traditional Indian medicine in practice in an Indian metropolitan city', *Social Science and Medicine* 15D: 69–81.

World Health Organization (1978) *The Promotion and Development of Traditional Medicine*, Technical Report Series No. 622, Geneva: WHO.

—— (1984) *Report of the Working Group on the Integration of Traditional Medicine in Primary Health Care*, Manila: WHO Regional Office for the Western Pacific.

—— (1985) 'Report of the consultation on approaches for policy development for traditional health practitioners, including traditional birth attendants', unpublished document, NUR/TRM/85-1, WHO Regional Office for South-East Asia, New Delhi.

5

CULTURAL AND DEVELOPMENTAL FACTORS UNDERLYING THE GLOBAL PATTERN OF THE TRANSMISSION OF HIV/AIDS

Nicholas Ford

INTRODUCTION

Since the first description in the medical literature in 1981 of the syndrome later described as *acquired immune deficiency syndrome* (AIDS), there has been a growing awareness that it represents one of the greatest threats to public health of the late twentieth century and beyond. By 1992 the World Health Organization estimated that there had been nearly 1.5 million cases of AIDS, as well as between 9 million and 11 million cases of *human immunodeficiency virus* (HIV) infection in adults and 1 million births with HIV worldwide (WHO 1992). These figures represent just the beginning of a growing global pandemic undermining progress in mortality and morbidity levels and reflecting enormous and deepening human misery. This chapter considers, first, the HIV/AIDS pandemic, its epidemiology and regional variations; secondly, reference is made to the cultural and developmental factors which help account for the regional variations in the level and pace of HIV/AIDS transmission; and thirdly, given that there is at present no cure for AIDS, nor vaccine to combat HIV infection, reference is made to the broad lines of strategies which are being developed to prevent the transmission of HIV. The focus throughout is upon transmission rather than the impact of the epidemic and efforts to cope with it.

At the very outset it is important to stress that in dealing with HIV/AIDS we are dealing with a social, as much as a biological and medical, phenomenon. HIV/AIDS is basically another new sexually transmitted disease (STD), and such diseases, being associated with the most intimate sexual behaviour, have for centuries of human history been subject to the taboo and social stigma through which societies have sought to regulate and control patterns of sexual behaviour (Selvin 1984). Thus with its combination of sexuality and

PERSPECTIVES AND ISSUES

lethality the emotive subject of HIV/AIDS infection has often encountered the responses of the so-called 'third epidemic' of denial, blame, stigmatization, prejudice and discrimination (Panos Institute 1990). The social challenge of HIV/AIDS is to foster patterns of behaviour, conducive to preventing transmission, and to be capable of responding compassionately and without discrimination towards those infected.

GLOBAL EPIDEMIOLOGY

A crucial factor in the development of AIDS is that following an HIV infection there is an incubation period of variable duration, during which time the body's immune system is progressively attacked, creating increasing vulnerability to a range of opportunistic infections (including, among others, pneumonia, tuberculosis, thrush, herpes, chronic diarrhoea, Kaposi's sarcoma, blood disorders and mental disorders). It has been suggested that the asymptomatic incubation period can last up to twenty years, but once AIDS develops it appears to be fatal (Panos Institute 1990).

The three main means of HIV transmission are from the exchange of body fluids in sexual intercourse (with greater biologic efficiency of transmission from male to female than vice versa), through infected blood (primarily via sharing needles/syringes and via blood transfusions) and perinatally from mother to child during pregnancy or birth. The dominant forms of transmission are found to vary widely between different regions.

It appears that prior to the clinical diagnosis of AIDS in 1981 transmission of HIV goes back to the 1960s and probably the 1950s. Questions of the geographical origins of HIV are futile and sterile; and, given their association with blame and stigmatization, they are here ignored. The crucial issue is to be able to understand the behaviours underlying transmission in order to develop and implement effective preventive strategies.

Given the variable and lengthy HIV incubation period, it is more useful to focus upon levels of HIV infection rather than on full-blown AIDS to understand the nature of the epidemic. Furthermore, in the light of the assured high levels of under-diagnosis, under-reporting and delays in reporting, it is important to work on the basis of (authoritative) *estimates* rather than actual reported cases (WHO 1992).

Figure 5.1 highlights the wide variations in the global distributions of HIV infections and their male–female proportions. This section will briefly review the main features of the HIV/AIDS epidemics in: sub-Saharan Africa; the 'Western' countries of North America, Western Europe and Australasia; Latin America and the Caribbean; and South and South-east Asia. These four groups of regions contain the vast majority of reported and estimated cases of HIV/AIDS.

As of 1992 sub-Saharan Africa is considered to have the highest numbers of HIV/AIDS cases of any major region, conservatively estimated at

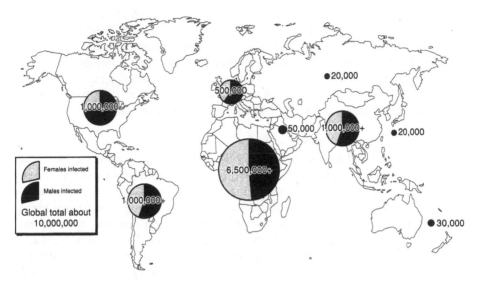

Figure 5.1 Estimated cumulative global distribution of adult HIV infections, January
1992
Source: Based on WHO (1992: 15)

cumulatively more than 6.5 million. Most of these cases are found in East
and Central Africa, but HIV/AIDS is also moving increasingly into Southern
and, to a lesser extent, West Africa. A second strain of HIV, HIV-2, has
been identified in West Africa, with recent research suggesting that it is
slightly less virulent than the HIV-1 strain found elsewhere in the world.
The dominant mode of transmission in Africa is by heterosexual intercourse,
with the male to female ratio being 1:1.2. The high level of other STDs,
especially those involving genital ulceration, has been considered an important
factor exacerbating the transmission of HIV in the region. HIV infection
from contaminated blood in transfusions and health-related infections is a
minor and declining source of infection, as blood screening and other health
service measures are expanded. Given the high level of infection of females
there is an increasing problem of perinatal transmission of HIV.

As of 1992 the affluent, industrialized countries of North America,
Western Europe and Australasia accounted for an estimated 1.5 million cases
of HIV infections, with approximately two-thirds of those considered to be
in the United States. The predominant population groups affected by the
epidemic are homosexual and bisexual men and injecting drug users, but
heterosexual infection is generally believed to be increasing. Although
homosexual/bisexual men comprise the overall majority of HIV cases, the
proportion of injecting drug users is higher in some areas (e.g. Spain, Italy)
than others. Furthermore, there is evidence that, while the pace of transmission among homosexual and bisexual men has declined, in response to

widespread behaviour changes, there are still rapid rates of infection among injecting drug users in areas where they continue to share needles/syringes. Perinatal infection is increasing as more women become infected with HIV.

In Latin America and the Caribbean the cumulative number of HIV infections is estimated to be over 1 million. In this region this rapidly spreading epidemic initially involved homosexual and bisexual men, followed by injecting drug users, but today includes increasing levels of heterosexual and perinatal transmission.

South and South-east Asia witnessed a somewhat later onset of the HIV/AIDS epidemic, but one which in places has involved some of the most explosive rates of increase. As of 1992 it is considered that there have been in excess of 1 million HIV infections, in particular in India and Thailand. Following relatively small numbers of early cases among homosexual men, the initial large-scale epidemic was observed among injecting drug users. Today the predominant mode of transmission in Asia is heterosexual, often between male customers and female prostitutes, and subsequently to customers' wives. This has, in turn, led to growing rates of perinatal transmission.

While the question of the impact(s) of the epidemic is, strictly speaking, beyond the scope of this chapter, some broad assessment can be made. First, the major impact of the epidemic will be increasingly felt in the future as large numbers of asymptomatic HIV cases develop full-blown AIDS. In all regions the loss of productive and creative labour will have enormous social and economic impacts upon societal functioning and development. In particular in those areas where transmission is primarily heterosexual, and large numbers of women are infected, there is a deepening problem of orphanhood. It is estimated that 70 per cent of children born to HIV-infected mothers will not in fact be infected, thus increasing the numbers of orphans as their mothers develop AIDS. The economic costs of treating and coping with AIDS are enormous and well beyond the health budgets of less developed countries (LDCs). In areas, such as parts of Uganda, which have already been hard-hit by AIDS morbidity and mortality, communities and families have had to respond to the losses. Barnett and Blaikie (1992) have documented, for instance, changes in patterns of agriculture as orphaned families struggle to survive. It is likely that a wide range of changes in community and family structures will take place as different societies seek to cope with the impact of AIDS on their populations.

Returning to the principal focus of this chapter – the transmission of HIV – it is clear that there are shifting patterns and rates of infection across different regions. In general terms the pace of transmission since the mid-1980s has been somewhat slower than was expected in the Western affluent countries, but more rapid than was expected in some regions of the less developed world.

The precise factors underlying the likelihood and speed of HIV transmission are still not fully understood, although progress is continuing to be made towards this end in basic scientific and epidemiological research. Basic insight into the epidemiological principles underlying HIV transmission has been derived from prior research into other sexually transmitted, persistent infections, notably herpes simplex (Nahmias *et al.* 1988). Two major factors, 'the pool' and 'poly-partnerism', are seen to influence the likelihood and pace of transmission. The pool comprises the total availability of the transmissible and infectious agent within a population – in short, the proportion of the population already infected. In the case of HIV, poly-partnerism is the number of different contacts with whom an individual engages in unprotected coitus or in needle/syringe sharing. Epidemiologically the prevalence of a sexually transmissible infection is primarily a result of the interaction of these two factors. The 'fullness' of the pool will effect the likelihood of infection, while the degree of poly-partnerism will determine how rapidly the number of infections will increase. The local and regional patterning of these epidemiological factors is shaped by issues of culture and development, to which the next section is addressed.

SEXUAL INTERACTION: CULTURE AND DEVELOPMENT

Given that HIV is primarily transmitted via sexual interaction, it is important to consider, first, the nature of sexuality, its psychosocial development and cultural context; and secondly, the ways in which aspects of social and economic development have an impact upon sexual interaction.

Sexuality can be defined as behaviour and social pathways which pertain to erotic stimulation, arousal and orgasm. Fundamentally, although sexuality has a biological basis, the form it takes depends upon socio-cultural patterning (Gagnon and Simon 1973). Human sexuality develops over the life course in a series of stages relating to physiological and psychological maturation and relationship formation. Throughout human history societies have sought to regulate and control human sexuality through systems of social sanctions and expectations. Sexuality is something which people learn and into which they are socialized. Thus there are variations in patterns of sexual interaction and expectations between different cultures. Furthermore, there are differences between social groups and individuals within specific cultures, which relate to differences in personality, life experience and (parental, peer-group and educational) socialization. Individuals appropriate and internalize particular socio-cultural norms regarding sexuality according to their own experience and preferences.

There are broad differences in the patterns of sexual interaction within different culture regions in terms of such indicators as the relationship contexts (before/within/outside marriage) in which intercourse takes place, the ages of first intercourse, the numbers of sexual partners and use of contraception and prophylaxis. These comprise important parameters which help to explain the pace of HIV transmission within specific societies once the virus has entered the pool. Thus, for instance, HIV transmission has been much faster in societies where there are high levels of unprotected sexual interaction with prostitutes, who have very many sexual partners, than in societies where the majority of people have few or moderate numbers of partners in their lifetime.

Given that sexuality is part of social behaviour, it is dynamic and influenced by broad social changes. For instance, in the Western industrialized affluent countries in recent decades there have been changes in patterns of sexual behaviour associated with the so-called 'sexual revolution', linked to social liberalization and technological advances in contraceptive technology. Similarly, in parts of Africa, Asia and Latin America rapid social change may be expected to have had an impact upon patterns of sexual interaction.

It is important to stress that specific impacts upon sexual interaction will depend upon the nature of the existing sexual culture. However, the major impacts of 'development' upon sexual interaction can be considered in terms of socio-cultural change, urbanization, mobility and the especial social dislocation of armed conflict. In many countries modernization and urbanization have brought in their wake new patterns of values, aspirations and lifestyles. Such socio-cultural impacts are probably most strongly experienced by the young. For instance, in parts of East Asia there are indications of increasing levels of pre-marital intercourse for both females and males as the young move towards perhaps more 'open', 'Western' patterns of sexual interaction. In sub-Saharan Africa there were wide variations between traditional ethnic groups in attitudes and values pertaining to pre-marital sexual interaction. It is often held that one of the effects of modernization in Africa has been the dismantling of many of the traditional constraints on pre-marital sexual activity. Mass communication and urbanization have been crucial in disseminating new 'modern' values and aspirations.

Where development increases the social and regional inequalities in economic opportunity and wealth, it may have an impact upon sexual interaction by increasing the economic attractions of prostitution. However, the social origins of prostitution are complex, as discussed with reference to the Thai case-study below.

A major feature of development has been increasing rates of mobility and migration at both international and internal levels. Mobility has two potential impacts upon sexuality: first, by bringing individuals and social groups into contact with different socio-sexual values (for instance, in the 'urban melting-pot'); and secondly, and more directly, by expanding the social and spatial parameters within which sexual interaction can take place. International tourism, for instance, can involve sexual interaction between tourists and host communities, with obvious implications for the introduction and transmission of STDs including HIV. The high levels of mobility (especially in circular migration) between the rural areas and growing urban centres in many developing countries provide a basis for widespread transmission of STDs. Where circular migration involves separation of partners, it can lead to increased recourse being taken by males to prostitutes. Also in some settings single females from rural areas can be vulnerable to sexual exploitation when they migrate to large urban centres.

As with other infectious diseases, the nodes of transport networks can be important for the transmission of HIV. For instance, the main transport routes have been viewed as facilitating the spread of HIV in parts of East and Central Africa. Furthermore, particular groups (such as truck drivers in East Africa and Thailand), who by nature of their occupation are particularly mobile, have disposable income and procure commercial sex, have been implicated in the regional spread of HIV.

Throughout human history wars and civil conflict have been recognized as important factors which can exacerbate the transmission of STDs. For instance, it was in response to the debilitating effects of STDs on the armed forces during the Second World War that some of the major modern advances in the development of anti-microbial and anti-bacterial treatments for STDs took place (Berg 1984). Armed conflict can effect STD transmission not only in terms of the socio-sexual activities of the armed forces, but also in terms of the social dislocation, hardship and marginalization of populations which it can engender. It must be stressed, however, that the specific ways in which patterns of economic development, mobility and conflict impact upon sexual interaction and HIV transmission very much depend upon a range of social characteristics of particular localities.

It is to the way in which local conditions and socio-cultural context relate to HIV transmission in one specific case-study, Thailand, that the next section is addressed.

CASE-STUDY: HIV TRANSMISSION IN THAILAND

This overview of the pattern and nature of HIV transmission in Thailand is structured with reference, first, to the emergence of HIV infection and the perception of it as a public health threat and, secondly, to some of the key socio-cultural factors (sexual culture, commercial sex industry, international tourism and drug dependency) implicated in the increasing transmission of the virus. In terms of the history of the HIV epidemic, Asia as a whole lagged behind other regions; however, it has long been recognized that the large scale of Thailand's commercial sex industry and the problems of injecting drug dependency placed that country in a position of particular vulnerability to HIV/AIDS.

In response to the diagnosis of the first AIDS cases (from 1984) in Thailand the public's reaction was largely one of complacency, with AIDS regarded as a homosexual problem largely confined to foreigners. It has also been suggested that the Thai government was hesitant in publicizing the growing threat of HIV/AIDS for fear of harming the country's burgeoning tourist industry. However, by 1987, facing pressure from both the medical establishment and the press, the government released figures and information concerning the level of AIDS in Thailand (Cohen 1988). Although the number of cases reported has risen in line with increased blood testing for HIV, it does seem likely that in the mid-1980s there probably were fairly few cases. As the number of cases in the 'pool' started to increase in the late 1980s, there was a huge increase in HIV transmission, until by late 1991 the Thai government estimated that there were between 200,000 and 400,000 cases of HIV infection. Furthermore, the government projections also suggested that this could increase to a cumulative total of 2 to 4 million HIV cases in Thailand by the end of the century (WHO 1992). HIV infection has now been detected in Thailand among injecting drug users (overwhelmingly males), prostitutes (both males and females), customers of prostitutes (males), recipients of contaminated blood products, wives of prostitutes' customers and, since 1989, babies born to wives of injecting drug users and customers of prostitutes. Within the range of sources of infection the greatest societal concern has surrounded infection from prostitutes to their male customers. This mode of infection is generally viewed as having the greatest potential for the widespread transmission of HIV to all strata of society.

One of the key features underlying Thailand's sexual culture is the strong double standard (inherent to traditional Asian sexual cultures) pertaining to the sexual activity of males and females. While it is often expected that young males may engage in pre-marital sexual intercourse, and married males may engage in extramarital intercourse, young women are expected to preserve their pre-marital virginity and wives to remain exclusively faithful to their husbands. A substantial component of female prostitution has developed to meet the demands of the double standard (Ford and Koetsawang 1991). Socialization into gender roles takes place in all societies. In Thai society boys can be more independent and wilful, while girls are expected to be more obedient and dutiful to parents.

In recent years there are some indications (survey findings and increased unwanted pregnancies among adolescent females not involved in commercial sex) that there is an increasing level of non-commercial pre-marital sex. This trend may reflect some aspect of liberalization/modernization, but it is important to stress that pre-marital sex is still strongly socially unacceptable for 'respectable' females. Given that there is little indication of any decline in the scale of the sex industry, it is likely that such non-commercial sex interaction is additional, rather than alternative, to recourse to commercial sex. Also, given that condom use is generally practised less in intercourse with girlfriends as opposed to that with prostitutes, sexually active young women are in a position of some vulnerability.

A brief mention will be made of the scale and diversity of Thailand's commercial sex industry prior to discussing its social context. There is continual recruitment of large numbers of young women into the industry. It is likely that there are between 200,000 and 300,000 women, mainly in the 16–24 age group, working in the industry throughout the country.

As in other countries, there is a wide range of commercial sex establishments and outlets, ranging through high-income 'call-girl' agencies, executive clubs, 'go-go' bars, coffee shops/tea houses, massage parlours, brothels (of different income levels), 'street-walkers' and even mobile commercial sex workers (CSWs) catering to rural market fairs. Although there is also a smaller sector of male prostitution catering to homosexual men, primary reference in this chapter is made to female prostitution. Involvement in the commercial sex industry is characterized by a high degree of geographical and occupational mobility, as workers seek to maximize their income potential. It should also be noted that, although part of the sex industry caters to foreign tourists and other visitors, this is a relatively small part of the whole.

The involvement of large numbers of females in the sex industry can be considered at two levels, a macro-level concerning the social and regional pattern of inequalities in wealth, and a micro-level concerning economic livelihood opportunities. Despite Thailand's remarkable rates of economic growth from the 1960s onwards, the country still remains predominantly (75 per cent) rural, dominated by the economic and political primacy of the Bangkok Metropolitan Region (BMR). The wide social and regional inequalities within Thailand are reflected in the large-scale migration (often circular and temporary) from the rural regions to the BMR, of which migration to work in the sex industry is a (small) part.

Although in absolute terms poverty in Thailand has been reduced over the past three decades, the increasing commodification/monetization of Thai rural society has served to increase the financial pressures on poor people and possibly also to make them more conscious of their relative poverty. It is within this perceptual and aspirational context that the economic attractions of the commercial sex industry need to be considered. Furthermore, a young woman's initial involvement in the industry is often related to family and community considerations. The young rural Thai woman is conscious of the need not only to support herself economically but also to make a contribution to her family's welfare. Helping the family through remittance of income from sex work is part of many daughters' dutiful relationship to their parents. Thus many CSWs from the rural areas remit funds back to their parents (who may or may not be aware of the precise origins of the income).

Without doubt the remittances to rural villages from daughters' work in the sex industry comprise a very substantial (although impossible to quantify) flow of capital. The developmental impact of this flow of capital largely depends upon how far it has been used merely for family consumption or for capital investments in economically gainful activities. It is likely that different CSWs and their families have made different uses of this income.

Social stigmatization and community social sanctions may be expected to discourage young women from working in the sex industry. However, the very scale of the numbers involved and the importance of the remittances have combined to reduce such sanctions in some communities. This is especially the case in parts of the northern region, where family pressure gives young girls little choice but to work in the sex industry, often following in their sisters' footsteps. It should be noted, however, that community attitudes to daughters' involvement in prostitution are different in other regions of Thailand – in the north-east it may be reluctantly tolerated as a means of coping with family poverty, while in the south strong negative sanctions militate against sex work. There are some indications that increasing numbers of young women are being recruited from the hill tribes in the north and from neighbouring Burma (Myanmar). The precise reasons for this trend are unclear, but may well relate to the potential for economic exploitation of these relatively naïve and marginalized groups. In Burma's case the general repression and lack of economic opportunity may serve as 'push' factors.

Serological testing in the sex industry revealed the first high rates of HIV infection in Chiang Mai in the north. The mobility of prostitutes and their customers has served to transmit HIV throughout the country. Serological surveys have also indicated much higher rates of HIV infection in the lower- rather than higher-income establishments. For instance, a comparative study of a high-income massage parlour and a low-income brothel in Bangkok revealed HIV positive rates of 1.6 per cent and 42 per cent respectively (Koetsawang and Ford 1993). Lower levels of condom use in the lower-income establishments may partly account for these rising rates of infection, although non-use of condoms was also found in the massage parlour. It would appear that the substantial minority of men who habitually refuse to use condoms in low-income brothels are continuing to infect large numbers of prostitutes. The basic problem is that prostitutes are currently unable to insist on condom use with customers.

Although the greatest social concern regarding HIV in Thailand has surrounded transmission via CSWs, the earliest high rates of infection were detected among a different sub-group, notably injecting drug users. Opiate consumption has a long history in South-east Asia, with considerable cultivation of the opium poppy taking place in the so-called 'Golden Triangle' which straddles northern Burma, Thailand and Laos. However, the processing of opium into heroin and injection as a mode of use are comparatively recent developments of the past three decades. Opium cultivation within Thailand itself has been reduced in response to the twin operations of police enforcement and development initiatives which seek to promote alternative (non-drug) means of livelihood for the hill tribe people. Meanwhile opium cultivation has continued to expand in neighbouring Burma and possibly also in Laos. Opium is grown almost entirely in upland peripheral areas of political instability and lack of government control (Lim 1984). It is likely that all of the main warring factions in Burma are using revenue from opium cultivation and heroin production to finance their military operations. The state terrorism of the Burmese military government and the consequent social dislocation and development collapse are probably major factors fostering the continued reliance upon opium cultivation in northern Burma (Silverstein 1990). Much of the heroin produced is clandestinely exported out of Thailand. The main social impact of the heroin trade upon Thailand has been that the very availability and accessibility of the drug have resulted in a deepening problem of heroin dependency. The high level of syringe/needle sharing among heroin users in Thailand resulted in very rapid rates of HIV infection, with over 40 per cent HIV seropositivity among injecting drug users attending clinics in Bangkok in 1988. The problem is being further exacerbated by opium dependants in northern Thailand currently switching from smoking to injecting drugs, creating the potential for a new HIV epidemic in that region.

Thailand is therefore at present a country with basically two HIV epidemics, one transmitted sexually and the other by needle/syringe sharing in drug dependency. Evidence for two epidemics has recently been provided by virological analysis which has revealed two basic sub-types of HIV-1 in Thailand (Ou et al. 1992). The sub-type found in female prostitutes, their customers and the pregnant wives of customers closely resembles an HIV sub-type found in Central Africa. The sub-type found in injecting drug users is essentially the same as that found in North American and European strains. Further research is seeking to understand the reasons for the apparent segregation between the two sub-types.

HIV PREVENTION STRATEGIES

Throughout the world HIV prevention strategies have contained different components seeking to reduce potential infection from the range of different potential sources. Health service efforts have been made to screen blood products to prevent HIV infection from blood transfusion, and to encourage adequate precautions in health service provision. Even these efforts are beyond the budgetary limitations of many developing countries, without considerable assistance from international agencies. However, the main HIV preventive efforts are concerned with minimizing the sexual and drug-using behaviours associated with transmission of the virus. The chief tools have been public and school-based strategies to educate people about the nature of HIV/AIDS and the means by which infection can be avoided. A further important component of AIDS education has been concerned with combating discrimination and stigmatization of HIV and AIDS sufferers. In many countries there has been debate concerning the content of HIV prevention education. One stance, often supported by politically and religiously conservative forces, has perhaps somewhat moralistically called for an attack on sexual promiscuity and drug use. An alternative, and perhaps less judgemental, stance has emphasized the need for consistent 'safer' practices, including condom use in intercourse and the avoidance of needle/syringe sharing in drug use. Interactive strategies involving outreach work to specific groups has a valuable role in seeking to influence behaviour towards safer practices. Discriminative policies which seek to punish and restrict specific groups (such as HIV-positive prostitutes and drug users) make it impossible to undertake interactive outreach work, as such groups are driven underground. Obviously the specific content and approach of any HIV prevention strategy have to be tailored to the specific sexual and drug-using culture in which it is being implemented.

From the foregoing discussion of the Thai context it is clear that HIV-preventive strategies have to go beyond general public and school-based education, to specifically and urgently address practices within the sex industry and the subculture of drug dependency. Indeed, merely *informing* the public about the risk of HIV/AIDS and the need for preventive behaviours is unlikely to have a significant impact upon behaviour.

The increase in condom sales and the reduction in levels of STDs in Thailand in the late 1980s are indicative of an increase in the level of condom use within commercial sex. However, this increase in condom use has not been sufficient to prevent an accelerating HIV epidemic. Although knowledge- and skill-based AIDS-related education for CSWs is vital, research shows that the crucial problem is that most CSWs are unable to insist on condom use with their customers. This highlights the pressing need for the implementation of the '100 per cent condom use in the sex industry' policy – giving the explicit right (even duty) to CSWs to refuse unprotected intercourse with customers. Given that the sexual interaction taking place is a private encounter between two individuals, the consistent and universal implementation of such a policy is obviously no simple task. Nevertheless, some sex establishments have already effectively implemented such a strategy; the need is to extend it to the whole country. It is crucial that the policy is explained to the proprietors of sex establishments, who have such an important role in its operation. Furthermore, it has to be universally applied, given the reluctance of some customers to comply with condom use and the commercial competition between different establishments (non-compliant customers will otherwise simply go to those places which do not insist on condom use). Such a policy has to be explicitly communicated to all customers both within sex establishments and through the mass media.

Careful socio-behavioural research needs to be addressed to helping to influence the behaviour of the non-condom-using customers.

In the longer term, it appears that the emergence of HIV/AIDS is fostering a wider social debate in Thailand about the nature of prostitution. Programmes which seek to reach and educate the communities and families which send their daughters to work in the sex industry have an important role to play. Obviously, the likelihood of HIV infection of the daughter should influence the cost-benefit calculus of the decision to send her to work in the sex industry.

In terms of reducing HIV transmission via injecting drug use, experience in several Western countries has highlighted the value of needle/syringe exchange schemes. However, in several South-east Asian countries there is still greater emphasis on preventing drug use *per se* rather than seeking to reduce the consequential transmission of HIV. Although Thailand is by no means as 'hard-line' in its anti-drug-use policies as neighbouring countries such as China and Malaysia, such thinking nevertheless has prevented the operation of large-scale needle/syringe exchange programmes. Education (via treatment/rehabilitation programmes) on the need to avoid needle/syringe sharing probably has made some impact upon injecting drug users in Thailand. The crucial problem remains, however, the widespread availability of heroin within Thailand, which is in turn linked to the enormous production in neighbouring Burma. In the long term, solutions to the heroin problem in Thailand are related to the seemingly insoluble political and military conflict in Burma.

CONCLUSION

This chapter has attempted to illustrate and emphasize the global variations in the scale and nature of the HIV/AIDS epidemic(s). The underlying theme has been that the form the epidemic has taken in the past, and is taking now and will continue to take in the future, depends upon regional and local historical, cultural and developmental conditions. The chief factors – notably the sexual and drug-using cultures – influencing the form of the epidemic were illustrated by reference to a single case, Thailand. It is only from a sound understanding of such dynamic processes that potentially effective HIV-preventive strategies can be developed. Ideally, HIV prevention should be approached within comprehensive, multifaceted and multisectoral strategies. Such strategies need to integrate communication and service-provision activities at different national, regional and local scales, and also to address short-, medium- and long-term goals. Finally, although the nature of the HIV epidemic varies across different regions and societies, it is vital that it continues to be viewed from a global perspective. Through the processes of communication, mobility and urbanization there are ever-increasing levels of international awareness, contact and interaction. HIV/

AIDS poses an enormous challenge to societies, their health and their developmental and political systems. HIV-preventive strategies need to be developed and implemented in ways which respect the human rights which *should* be at the core of the world order.

REFERENCES

Barnett, T. and Blaikie, P. (1992) *AIDS in Africa: Its Present and Future Impact*, London: Belhaven.

Berg, S.W. (1984) 'Sexually transmitted diseases in the military', in K.K. Holmes *et al.* (eds) *Sexually Transmitted Diseases*, London: MacGraw-Hill, pp. 90–9.

Cohen, E. (1988) 'Tourism and AIDS in Thailand', *Annals of Tourism Research* 15: 467–88.

Ford, N.J. and Koetsawang, S. (1991) 'The socio-cultural context of the transmission of HIV in Thailand', *Social Science and Medicine* 33(4): 405–14.

Gagnon, J.H. and Simon, W. (1973) *Sexual Conduct*, Chicago: Aldine.

Koetsawang, S. and Ford, N.J. (1993) *An Investigation into the Psychosocial Factors Influencing Condom Use by Female Prostitutes in Thailand*, Bangkok: Siriraj Hospital, Mahidol University.

Lim, J.J. (1984) *Territorial Power Domains, Southeast Asia and China: The Geo-Strategy of an Overarching Massif*, Canberra: Strategic and Defence Studies Centre, Australian National University.

Nahmias, A.J., Lee, F. and Danielsson, D. (1988) 'Epidemiological principles for understanding the prevalence of HIV infections and their possible control', in R.F. Schinazi and A.J. Nahmias (eds) *AIDS in Children, Adolescents and Heterosexual Adults*, New York: Elsevier, pp. 154–9.

Ou, C.Y., Auwanit, W., Pau, C.P. *et al.* (1992) 'Identification of two distinct HIV-1 sub-types in Thailand with apparent segregation by mode of transmission', paper for the Eighth International Conference on Aids, We C 1025, Amsterdam.

Panos Institute (1990) *The Third Epidemic*, London: Panos Institute.

Selvin, M. (1984) 'Changing medical and societal attitudes towards sexually transmitted diseases: an historical overview', in K.K. Holmes *et al.* (eds) *Sexually Transmitted Diseases*, London: MacGraw-Hill, pp. 3–19.

Silverstein, J. (1990) 'Civil war and rebellion in Burma', *Journal of Southeast Asian Studies* 21: 114–34.

World Health Organization (1992) *Current and Future Dimensions of the HIV/AIDS Pandemic: A Capsule Summary*, Geneva: WHO.

6

THE GLOBAL PHARMACEUTICAL INDUSTRY

Health, development and business

Wil Gesler

INTRODUCTION

The use of legally purchased drugs is increasing in the developing world at an alarming rate. Traditional healers have used drugs since time immemorial, but the new influx is largely created by the process of medicalization, or the tendency for biomedicine to entice people with the lure of scientifically based medicine. Drugs used for medical cures are a mixed blessing. They are essential to the treatment of many diseases prevalent in developing countries. If used properly, they can bring about enormous improvements in health care. However, there are many problems connected with the control, distribution and cost of drugs in the Third World that people interested in health and development must be aware of. This chapter addresses the following topics related to drug use: (1) the ways in which drugs can be classified; (2) the market control exerted by multinational pharmaceutical companies; (3) the roles of national governments and international organizations; (4) personnel and facilities involved in the distribution of drugs; (5) major drug-related issues, including over-use and misuse, unhealthy and illegal practices, the role of the private and public sectors, drug costs, equity and system integration.

DRUG TYPOLOGIES

The proliferation of drugs makes it very difficult to understand their true value or how to control and allocate them. Some order might be brought to this confusion by pointing out several ways in which pharmaceuticals can be categorized. Note that the boundaries between the dichotomies described below are often quite fuzzy.

Brand names versus generics

The distinction between brand names and generics is important because multinational pharmaceutical companies, naturally enough, promote their own brands, claiming that, although their drugs may be more expensive, they carry a guarantee of good quality. Promotion of brand names has a very strong effect on doctors and pharmacists, who tend to prescribe brands over generics. The general public also tends to be swayed in the direction of brand names.

Prescription versus over-the-counter (OTC) drugs

Prescription drugs can by law only be sold by a health facility or pharmacy if the buyer has a physician's prescription, whereas OTC drugs carry no such restrictions. Even though more developed countries (MDCs) have economies dominated by the open market system, they usually have relatively strict controls over OTC sales. Less developed countries (LDCs) tend to be much more relaxed in this regard, and so many drugs requiring prescriptions in MDCs are sold OTC in LDCs. Furthermore, it is more common in LDCs to sell prescription drugs as OTC drugs illegally or out of ignorance. The dangers of selling too many OTC medicines are obvious. Many governments turn a blind eye to OTC sales because they lack the will or ability to police them.

Essential versus non-essential drugs

Essential drugs are those that medical authorities in a country deem to be priority drugs for them, given the prevalence and severity of diseases their populations experience and the cost-effectiveness of available treatments. The majority of drugs on the market today are non-essential by this definition, so governments which are serious about controlling the influx of pharmaceuticals have to make decisions based on this dichotomy.

Preventive versus curative drugs

The distinction between drugs which help prevent problems from arising and those that help cure problems after they have arisen is very vague, because illness and health are a continuum, and many drugs can be used for both purposes. However, it is important to think in terms of this dichotomy, because health planners have to make decisions about what proportion of their budgets should be spent on prevention or cure. In the long run, curative medicine is usually far more expensive than emphasis on prevention. Therefore, budgeting for high proportions of preventive drugs (vitamin supplements, malaria prophylaxis, immunizations for under-5s, and so on) is desirable.

Therapeutic versus convenience drugs

The therapeutic–convenience distinction is between those drugs that have been shown by clinical trials to be useful in preventing or treating certain conditions and those which are used to enhance people's enjoyment of life. The latter drugs, which include pain-killers, tonics and a range of stimulants, sedatives and tranquillizers, are easily available but have questionable long-term efficacy, often have very undesirable side-effects and can lead to drug dependency.

Traditional versus biomedical drugs

Biomedical drugs are an adjunct of biomedicine wherever it is practised. Traditional societies usually have a wide variety of medicines prepared from various plant and animal substances which are prescribed by traditional medical practitioners (TMPs) or used in self-medication. Research has found that many of the plant species used in traditional remedies contain chemical compounds which have curative power, and so plant extracts are very often used in biomedical preparations. Gathering the necessary raw materials and producing remedies from them is a very important occupation, and in some countries, such as China, India and Korea, traditional drugs are manufactured on an industrial scale. Traditional drugs are usually more easily obtained than modern pharmaceuticals. They are also very important because they are culturally acceptable and can have important psychological effects. They do have some drawbacks, however. Their efficacy for specific ailments is often unknown; their use is inconsistent; dosage is often haphazard; and they may cost more than biomedical drugs.

MULTINATIONAL PHARMACEUTICAL COMPANIES

An overriding factor in any discussion of drugs is the dominance of multinational pharmaceutical companies (MNPCs) over the world drug market. Over the last thirty years, MNPCs have been among the fastest-growing multinationals, with annual growth rates varying between 12 and 20 per cent. Worldwide sales for 1991 were projected to be around $120 billion. Large drug firms, which are based for the most part in MDCs, are very eager to supply pharmaceutical products to the rapidly expanding LDC market, which now demands about 25 per cent of their output.

Market control

MNPCs claim that they operate in an open market economy, in competition with each other, but in fact they constitute an oligopoly; that is, a very few companies control international production, trade and innovation. MNPCs are concentrated both structurally (the leading 100 firms produce about 70 per cent of all drugs) and geographically (in 1980 seven countries – the United States, Japan, West Germany, the United Kingdom, France, Italy and Switzerland – controlled 76 per cent of the market). It is not surprising that this oligopoly exists, because drug production and research are very capital-intensive enterprises. Successful companies can use their profits to corner raw materials markets and to continue to search for new formulations.

It is difficult for LDCs to break the market power of MNPCs because these firms, like all multinationals, operate on a worldwide basis. If a particular country shows dissatisfaction in its dealings with an MNPC, the company can take its business elsewhere or perhaps withdraw a subsidiary company from the recalcitrant country. MNPCs are well aware of the 'business climate' in each country. They assess the advantages and disadvantages of dealing with each place and act accordingly.

Control strategies

Large drug firms maintain control of markets in various ways. One major strategy is to mount expensive campaigns to market and promote their products. In these campaigns there is a strong tendency to grossly exaggerate the merits of individual products and to ignore or quickly pass over any dangers. Words such as 'power' and 'strength', for example, are used to indicate that a tonic will give a person virility. Whereas countries such as the United States have strict disclosure laws relating to pharmaceutical products, most LDCs do not. Drugs that can only be used for certain conditions in MDCs are marketed for a wide range of ailments in LDCs.

As a whole, MNPCs spend more money on advertising and trade promotion than they do on research (about 19 per cent as compared to about 10 per cent). Much of this money goes to hire 'detail men' who pressure physicians to buy drugs that may be inessential or inappropriate. A great deal of money is also spent on producing glossy brochures, free samples for doctors and pharmacists, scientific articles defending drug company positions on issues, and entertainment for clients. The International Federation of Pharmaceutical Manufacturers' Associations established a marketing code in 1981 to address marketing and promotion abuses, but sceptics view this as simply an admission of their guilt.

A second strategy used by MNPCs to control markets is research and development (R&D) activity. Searching for useful new drugs is an extremely expensive proposition, and only large firms have the capital which the average totally new product requires – $30 million to $60 million. Since individual governments simply do not have this kind of money, drug firms feel justified in monopolizing this vital aspect of the drug business.

MNPCs also invest heavily in branch plants in LDCs, a third control strategy. Many LDC governments welcome the investment and so they compete with each other to produce favourable business climates, which include little government interference in MNPC activities, cheap labour supplies and limited controls on prices charged. The problem is that the benefits of foreign investment mostly accrue to the large firms, and what money is retained within LDCs ends up in the pockets of a small group of government officials and business men.

ROLE OF NATIONAL GOVERNMENTS

The governments of LDCs, charged with the health of their populations, are struggling to come up with rational, cost-effective schemes that will supply appropriate drugs to everyone in need. As we have just seen, one of the major problems LDC governments face is control of markets by MNPCs. To break this hold, governments have attempted to limit prices on drug imports, restrict the range of drugs, encourage local production and establish state-owned companies to compete with private firms.

Strategies used to assert control over drug production and distribution vary widely among individual countries. Some governments are simply content to let the private sector handle the pharmaceutical sector of health care. Nigeria, Ghana and Sierra Leone, countries which have attempted to gain control of many aspects of business, have required foreign drug firms to invest capital within their countries, have set quotas on the number of foreign employees and have encouraged local participation in the drug industry. Indonesia, which is more economically independent than most LDCs, has hinted strongly that MNPCs must invest in local plants or risk exclusion from local markets. Pakistan has attempted to market only generic drugs, and Mexico has passed legislation to match the technological level of drugs to its local needs. Cuba, operating within its revolutionary political and social framework, centralized drug purchase and distribution, tried to achieve technological self-reliance and quality control, carried out its own R&D activities and made agreements with other socialist countries. In India, following a government report on the pharmaceutical situation, local drug industries were encouraged, prices were controlled, imports were constrained, and joint ventures with foreign firms were restricted to essential drugs.

Two major strategies have been employed by LDC governments: establishment of essential drug programmes (EDPs) and domestic drug production. A country embarking on an EDP must decide which drugs are essential by considering what diseases are prevalent, how severe they are and how cost-effective various drug alternatives might be. Since only fifty to sixty drugs are really required to treat about 80 per cent of a country's health problems, the vast majority of the approximately 30,000 drugs available on the world market must be severely pared down. In the late 1970s the World Health Organization drew up a list of 200 essential drugs and in 1981 it introduced a special programme on essential drugs and vaccines. Generics are usually preferred to brand-name drugs because they are cheaper, but care must be taken over their quality. Drugs which are known to be ineffective or harmful, and most convenience drugs, should be eliminated. Care must be taken to provide an ample proportion of preventive as opposed to curative drugs.

Despite the fact that, by 1987, over 100 countries had prepared essential drug lists, their implementation has encountered some serious difficulties. As expected, MNPCs and local wholesalers and retailers have resisted EDPs. LDC governments often lack the will to maintain these programmes and in some cases have even obstructed them. There are also organizational problems of inadequate facilities for storage, transport, packaging and labelling, quality control and inventory control. However, some countries have had limited success. Mozambique reduced a list of around 26,000 available drugs to 120 and registered these for sale in pharmacies.

Several countries have been fairly successful in establishing their own drug industries. Brazil and Egypt both produce most of their own drugs locally, and in Bangladesh 90 per cent of the drugs which are sold have been manufactured within the country. India is producing drugs for its own consumption and is even exporting some products. Domestic drug production faces several barriers, however. Locally produced drugs may not be cheaper than imports, given the requirements of capital investment and economies of scale, and quality control may be poor. Production workers must be trained, and the necessary technological skills acquired; and, again, resistance from MNPCs will be encountered.

ROLE OF SUPRANATIONAL ORGANIZATIONS

Since individual countries often have great difficulty in controlling drug markets, it would seem that regional or global strategies might be a solution. Some regional groupings do exist. The Pan-Arab Pharmaceutical Company established priority lists for each member country, developed national agencies to purchase drugs, revised patent and trade-mark laws, and established domestic industries. Another attempt at regional co-operation, this one by the Caribbean community, met with serious problems: not every country in the region joined, and members, with different political ideologies, economic priorities and social concerns, often could not agree on policies.

As mentioned above, international organizations such as the WHO have been actively involved in drug issues. The WHO's former director, Halfdan Mahler, attacked MNPC practices, but the agency usually has not been very aggressive. Other international organizations that have tried to regulate the international drug trade include the International Labour Organization, the International Confederation of Trade Unions, Health Action International, the United Nations Industrial Development Organization, the United

Nations Conference on Trade and Development and UNICEF (the United Nations Children's Fund). Unfortunately, although these organizations have formulated policies, they have accomplished little of a substantive nature. Their hands are often tied because they rely on MDCs, the home of the MNPCs, for financial support and on the drug companies themselves for technology. Furthermore, the international agencies often disagree among themselves over appropriate strategies.

DRUG DISTRIBUTION PERSONNEL AND FACILITIES

Drugs are likely to pass through many hands before they reach the consumer. Because there is such a complex chain of people and facilities, both public and private, involved, it is difficult for control to be maintained over the transfer and sale of pharmaceuticals. In this section we discuss some of the most important components of drug distribution.

Physicians and pharmacists are usually the main dispensers of drugs. The amount of training they have received in the proper use of medicines varies from place to place. They are under a great deal of pressure from pharmaceutical firms to prescribe a variety of products and they do not have time to read all the pertinent literature on new drugs. Physicians are key personnel because they have the most authority over drug prescription. Many doctors work in a public facility, such as a government hospital, part of the time and in a private practice at other times, but they prescribe drugs in both capacities. Many also own or are part-owners of pharmacies.

Pharmacists provide an important alternative to physicians, because they are often more accessible and many people trust their knowledge of drugs. In areas where few or no doctors are available to prescribe and dispense drugs, pharmacists may take on that role and even give out prescription drugs while the authorities look the other way. Some pharmacists 'diagnose' the diseases of their customers and prescribe accordingly. The dangers of incorrect diagnoses and prescriptions are obvious. However, with adequate training, pharmacists can play a very useful role in health care delivery.

There are basically three systems of facilities for drug distribution. National governments – the public sector – use their hospitals, health centres and dispensaries. The private sector dispenses drugs either through legal, licensed retailers in pharmacies and patent medicine stores or else through a very wide range of vendors who are unlicensed. The latter system, which may include sellers of small packets of pills and powders in a village market, itinerant drug pedlars and TMPs who sell Western as well as native remedies, is ubiquitous. Again, governments are faced with the problem of what to do with this informal part of the health care system which provides a service that is sometimes necessary but is often useless or dangerous.

MAJOR DRUG ISSUES

This section of the chapter summarizes several of the major issues that have arisen in drug production and distribution. Some important problems and possible solutions to them will be discussed.

Over-use and misuse

Many health planners make the mistaken assumption that more drugs and cheaper drugs are automatically better for the health of a population. Since drugs often act in a dramatic fashion, and because pharmaceuticals are so forcefully marketed and can so easily be used in an unthinking manner, there is a tremendous problem of their over-use in health care. Some people might even think that using drugs is all there is to health provision. We are speaking of over-use in general here, but specific products are also used too much.

Dispensing drugs often detracts from other important aspects of health care, such as proper nutrition, monitoring the progress of young children and clean water supplies. More attention may be paid to who prescribes drugs rather than to whether or not they should be prescribed at all. There seems to be a great deal of hypocrisy, in both MDCs and LDCs, over drug use and abuse – we deplore the taking of illegal drugs and conduct 'wars' against producers and dealers in illegal drugs, and yet overdose ourselves with legal products. Many medicines are prescribed or sold with little or no knowledge about their indications, dosages, efficacy or side-effects.

Unhealthy and illegal practices

Besides drug misuse and over-use, there are several activities that are illegal or that work against the proper use of drugs. Perhaps the most commonly known unhealthy practice is that of drug dumping. Products banned for sale in MDCs because of their dangerous qualities are sold to LDCs, whose regulations are far less strict. For example, the drug dipyrone was sold in Latin America, with no warnings on the label, for the treatment of minor ailments after it had been banned by the US Food and Drug Administration because it could cause a severe blood disorder.

Besides dumping drugs, MDCs and MNPCs also try to sell inappropriate pharmaceuticals. In 1980, only 1.3 per cent of North Yemen's drug imports were for treatment of three of its most prevalent diseases – malaria, schistosomiasis and tuberculosis – whereas 17.8 per cent of its imports were of tonics and vitamins. In their R&D activities, drug firms concentrate on problems prevalent in the West such as cancers and heart disease, while investigation of medicines for preventing and curing tropical diseases lags far behind (see Chapter 1 on the epidemiological transition).

Although we hear a great deal about the illegal drug trade (e.g. cocaine and heroin), not too much is said about illegal activities connected with the legal trade. However, there is bound to be theft and fraud in any business that makes so much money. Drug theft typically takes the form of siphoning off supplies from public-sector stocks and filtering them through to private vendors – the Kenyan government, for example, may have lost 40 per cent of its drugs in this manner. Other common practices include bribing corrupt officials to accept tender offers or making pay-offs to pharmacists who accept certain drugs for sale.

The role of the public and private sectors

The production and distribution of drugs are an excellent example of conflicting goals between the public and the private sectors. National and local governments, representing the public sector, are charged with providing best health care possible for their people, whereas private firms or individuals have as their primary goal the making of profits. Of course, the private sector may take humanitarian actions, and the public sector is often involved in profit-making schemes. None the less, the drug industry is overwhelmingly dominated by the private sector, with the result that economic concerns generally outweigh health concerns.

There are definite advantages to keeping drug production and distribution within the private sector. Pharmaceuticals are a thriving business, and large companies can be very efficiently run. MNPCs can help LDCs balance their foreign payments through investment in branch plants, help train an industrial workforce and foster other moves towards development. Drug firms have the expertise and infrastructure for the widespread distribution of their products. In many areas of LDCs, at present, private vendors are the only source or the most easily accessible source of drugs. It is not surprising that many LDC governments are willing to relinquish part of their health care responsibility to private concerns.

Arguments for more public-sector involvement centre around the idea that the health of all the people transcends concern for turning profits. True development, from this perspective, should not be measured solely in terms of economic indicators. Healthy, productive populations are more important in the long run, and indirectly lead to economic gains. Achieving social and cultural goals is just as important as raising the gross domestic product. The public sector is potentially in a much better position to distribute drugs on a socially and geographically equitable basis. Furthermore, as we have seen, much of the activity of the private drug sector is detrimental to health.

Drug costs

Drug purchases comprise a very large proportion of health budgets in developing countries. Whereas MDCs spend around four times the proportion of their gross domestic product on health care, the LDCs spend far greater proportions of their health care budget on drugs – about 40 per cent compared to about 8 per cent. What is perhaps even more alarming is that the use of drugs has been increasing faster in LDCs than in MDCs.

The cost of pharmaceuticals and who will pay these costs are another major issue. There is great concern that the prices set by MNPCs are excessively high and, in fact, that these firms engage in price fixing. Over the last thirty-five years, pharmaceuticals have consistently been the first or second most profitable manufacturing industry in the world. If prices were fairly set, they should be fairly uniform, but there is strong evidence that prices vary considerably from country to country, depending on what local markets will bear.

Drug prices remain high for several reasons. MNPCs, as an oligopoly, do not compete to the extent that they would on an open market, and when they do compete they tend to do so over products rather than prices. Drug firms justify their charges by the high costs of R&D, but much of this activity is unnecessary. Also, large sums are spent on marketing and promotion. Another cost-raising practice is transfer pricing, in which MNPCs charge LDC subsidiaries inflated prices for raw materials, machinery and technical assistance. Brand-name drugs, with the same efficacy as generic products, but which cost more, are promoted. Doctors who prescribe drugs without regard to their cost are also part of the problem.

Several suggestions for dealing with the price of drugs have been made. Probably the most widespread idea is to charge consumers for drugs as part of a 'cost recovery' solution to financial problems. In 1987 UNICEF proposed the Bamako Initiative in which people in poorly served rural areas would be sold drugs for two to three times their cost price and the profits would be used to set up a revolving fund to buy more drugs and also to support other local health care activities. The proposal faces many difficulties, including the inability of many poorer people to pay; conflict with government guarantees of free health provision; problems of setting equitable prices; declining moneys as the fund revolves; and problems in collecting debts when drugs are sold on credit.

Some 'two-tier' pricing solutions have also been put forward. In one, poorer nations would simply be required to pay less for drugs. In another, essential drugs would bear a lower price than the market price, and the difference would be made up by charging people more for non-essential and convenience products. Sending out tenders for bids by pharmaceutical companies could also cut down on costs. Sri Lanka tried this quite successfully. It was able to buy good-quality products at relatively low prices from relatively small firms.

106

Equity

It is very clear that, given the current situation, there will be gross inequities in the distribution of drugs. Inequities are both social and geographical. The public and private sectors tend to cater to the wealthy. Private vendors, in order to make higher profits, naturally tend to target those who can afford their products. One industry spokesman has gone so far as to state that MNPCs purposely cater to the middle- and upper-class 'productive' elements of society and he deplores the efforts of international agencies to aid the unproductive 'social periphery'. The public sector lacks the infrastructure to reach the poor, who often are forced to buy in the private sector. And, ironically, those who can least afford to buy drugs may spend proportionally more of their money set aside for health care on drugs.

Geographic inequities exist at different scales. At the grossest level, MDCs are at a great advantage compared to LDCs in terms of drug availability, appropriateness and quality. Within countries, rural areas tend to be peripheral to urban centres. A study in India, for example, showed that 40 per cent of the people who attended clinics and required drug treatment were sent away to buy the drugs elsewhere, because clinic stocks were depleted. Within towns and cities, urban élites have the advantage over have-nots.

System integration

Chapter 4 of this volume addresses the important question of the potential for incorporating traditional medicine into modern health care systems. We spoke earlier of drugs acting as a bridge between biomedicine and traditional medicine, tending to draw people to the latter from the former. Drugs are also potentially a means of helping health planners to integrate traditional and modern systems. After all, a great many people pragmatically use drugs produced by both systems anyway. If some of the remedies used by TMPs can be shown to be effective in preventing or treating certain conditions, then those practitioners and those drugs could be brought into a unified delivery system. As always, difficulties would be encountered. The quality and dosage of traditional remedies would have to be established and monitored; TMPs would have to be trained to administer their drugs more effectively; and biomedical personnel would have to make some attitude changes towards TMPs and traditional medicine.

FURTHER READING

Blum, R., Herxheimer, A., Stenzl, C. and Woodcock, J. (eds) (1981) *Pharmaceuticals and Health Policy*, London: Croom Helm.
Heller, T. (1977) *Poor Health, Rich Profits: Multinational Drug Companies and the Third World*, Nottingham: Spokesman.

Patel, S.J. (ed.) (1983) *Pharmaceuticals and Health in the Third World*, Oxford: Pergamon.

Pradhan, S. (1983) *International Pharmaceutical Marketing*, Westport, CT: Quorum Books.

Silverman, M. (1976) *The Drugging of the Americas*, Berkeley, CA: University of California Press.

Thrupp, L. (1984) 'Technology and planning in the Third World pharmaceutical sector: the Cuban and Caribbean Community Approaches', *International Journal of Health Services* 14(2): 189–216.

Tucker, D. (1984) *The World Health Market: Future of the Pharmaceutical Industry*, Guildford: Euromonitor.

World Bank (1993) *World Development Report 1993*, New York: Oxford University Press.

Part II

DEVELOPMENT AND
THE HEALTH OF
PLACES AND GROUPS

7

CITIES AND HEALTH IN THE THIRD WORLD

Trudy Harpham

INTRODUCTION

This chapter examines urbanization in developing countries and associated health problems. It can be argued that the most important philosophy guiding the development of health policy and planning in developing countries during the 1980s was primary health care (see Chapter 12). Most of the discussions and literature on primary health care have been related to rural areas of developing countries and have rarely addressed the problems of the rapidly growing cities. Typically up to 80 per cent of developing countries' national health budget is spent in cities. This is due to the major hospitals of the country being located in the cities and the priority given to, and expensive nature of, tertiary (hospital) care. When examining equity (distributive justice) in health care at a national level, it was often argued that rural areas deserved priority attention due to the inequity of this 'urban bias' (Lipton 1976). During recent years, however, more attention has been focused upon equity within Third World cities themselves. It has been recognized that there is an equally urgent need to develop primary health care within cities as there is in rural areas. The rapid rates of urbanization and the associated growth of poor urban populations have prompted health policy-makers and planners to raise questions about the appropriate way in which to develop urban primary health care. This chapter highlights the relevant demographic trends in this debate, describes the health problems that need to be tackled and summarizes recent research and policy directions.

URBANIZATION IN DEVELOPING COUNTRIES

Urbanization can be defined as the relative increase in the urban population as a proportion of the total population. Urban populations have greatly increased in the last forty years, and the greatest increase has been in the developing countries (Figure 7.1). Urban populations will increase even more rapidly in the next thirty-five years, with an explosive growth of cities in developing countries. Although the number of rural dwellers will

111

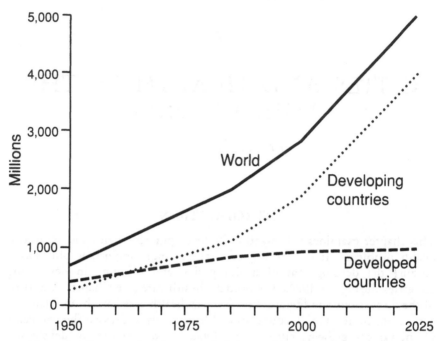

Figure 7.1 Increases in urban populations, world and developing countries,
1950–2025
Sources: UN Population Division (1984); UN Department of International Economic and
Social Affairs (1989)

increase, their share of the total population will not (Figure 7.2). People living in cities will become a global majority, and the developing countries will approach the urbanization levels of the industrialized countries. The spectacular growth of the mega-cities has received a great amount of attention. In 1950 there was one Third World city with a population of more than 5 million. There were eleven in 1970, and thirty-five are projected by the year 2000. Of these thirty-five cities, eleven are forecast to have a population of 20 to 30 million. Although such spectacular growth attracts much attention, it is important to appreciate that most urban dwellers will be living in smaller cities. Environmental and health problems in the mega-cities will be the more severe, but the needs of the remaining cities will be the more extensive.

In any debate on urbanization it is important to bear in mind certain problems about definitions of 'urban' and the methods used to form population projections. The definition of 'urban' varies from country to country. For example, in Peru 'urban centres' are those with 100 or more occupied dwellings; in India they are settlements with 5,000 or more inhabitants. Mostly, definitions fall between 1,500 and 5,000 people, but there are undoubtedly problems when comparing urban populations. It is

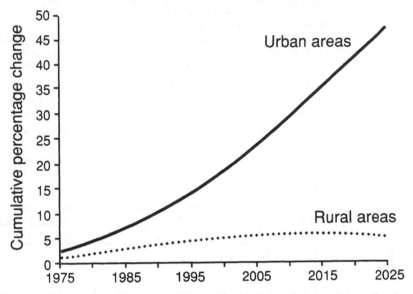

Figure 7.2 Percentage change in populations of urban and rural areas in developing
countries, 1975–2025 (base: 1970)
Source: WHO (1991: 4)

safer to examine the *number* of large cities for comparative purposes, since
the same criterion is used for each nation. But even here the statistics for
some countries are known to be inaccurate. Many countries have not had a
census for years, so extrapolations are from very old data. Another problem
to consider is that sections of poor urban populations may be regarded as
unofficial settlers and therefore not counted in censuses.

Contrary to what is often supposed, natural increase and not migration
is the major factor for urban population increases in the world. Globally,
natural increase is responsible for an average of 61 per cent of urban
population growth in developing countries, compared to only 39 per cent
from rural migration. Clearly, this pattern differs between continents and
countries, with natural increase being more important in most Latin
American countries, for example, while rural-to-urban migration is still the
main contributing factor to urbanization in most sub-Saharan African
countries. However, regardless of what policies are undertaken to affect
internal migration, and their successes or failures, governments have to come
to terms with the increasing numbers of the urban poor being born in the
cities.

It is not simply the absolute level of urbanization (i.e. the percentage of
the population living in urban areas) that is important, but also the relative
speed with which urbanization is occurring. Many Latin American countries
are very urbanized, with over 70 per cent of their population living in urban

113

Table 7.1 Differences in urbanization patterns of the developing countries

Group 1:	Heavily urbanized countries of more than 75 per cent urbanization, usually including mega-cities but with declining rates of urban growth. Most growth attributable to natural increase rather than migration. Typical of Latin America, e.g. Argentina, Mexico, Colombia, Brazil.
Group 2:	Recently urbanizing with about half the population living in urban areas. Growth rates have peaked and are beginning to decline. Typical of North Africa and some Asian countries, e.g. Algeria, Morocco, Malaysia.
Group 3:	Primarily rural but rapidly urbanizing. Migration a major source of urban growth, although male migration is being replaced by household migration, leading to a shift towards natural increase as major growth. Typical of African countries, e.g. Senegal, Ivory Coast, Nigeria, Sudan, Kenya, Zaire.
Group 4:	Large, mostly rural countries. Major urban concentrations. Urban growth rates stabilized at high levels and projected to continue for next decade. Typical of large Asian countries, e.g. India, Indonesia, China.

Source: adapted from World Bank (1991)

areas, but the most rapidly urbanizing countries tend to be in sub-Saharan Africa and Asia (Harpham *et al.* 1988). Table 7.1 presents a categorization of different patterns of urbanization in developing countries and illustrates the importance of the evolutionary stage of urbanization in any one country.

GROWTH OF THE URBAN POOR

However inaccurate projections are, there is no doubt that hundreds of millions of city dwellers are already exposed to major threats to their health and well-being. A major cause of this threat is poverty. Although comprehensive statistics on urban poverty are lacking and national criteria that define poverty differ, one can estimate that about one-third of urban dwellers in developing countries live in substandard housing or are homeless, and that their numbers will grow to at least 420 million by the year 2000. Although urban incomes are generally higher and urban services and facilities more accessible, poor town dwellers may suffer more than rural households from certain aspects of poverty. The urban poor, typically housed in slums and squatter settlements, often have to contend with appalling overcrowding, bad sanitation and contaminated water. The sites are often illegal and dangerous. Forcible eviction, floods and landslides, and chemical pollution are constant threats.

An extensive literature considers traditional and current explanations of both urbanization and urban poverty. These will not be discussed here. The health implications of such urbanization and urban poverty are discussed below.

URBAN POVERTY AND HEALTH PROBLEMS

In some developing cities, the poor make up as much as 60 per cent of the population, and the health problems of these urban poor are becoming more critical (Tabibzadeh *et al.* 1989). These health problems involve both diseases which are traditional to the developing countries as well as diseases which have been associated with higher levels of development and industrialization. The urban poor have been characterized as being at the interface between underdevelopment and industrialization. In other words, they are at a mid-stage in the epidemiological transition (see Chapter 1). High levels of traditional health problems are indicated by high maternal, perinatal, infant and child mortality rates and by malnutrition and infectious diseases. 'New' health problems associated with urbanization and industrialization include cancers, hypertension, mental illness, problems of drug and alcohol use, sexually transmitted diseases (including AIDS), accidents (traffic and industrial) and violence. With this double burden of health problems the urban poor may be characterized as suffering the 'worst of both worlds' (Harpham *et al.* 1988).

Measuring the extent of the health problems of the urban poor is difficult. If the data source is hospitals and health facilities there is the problem that often the urban poor do not use these facilities and are therefore not covered in such health statistics. If data come from community-based measures – for example, national health interview surveys – there is the problem that the urban poor might not be included in such official statistics, plus the additional problem of aggregated city health data, which mask the degree of health problems of the urban poor. For example, infant mortality (deaths by the age of 1 per 1,000 live births) might be around 80 in the city as a whole but reach 200 for certain slum areas.

The scarcity and contamination of water supplies and the lack of sanitation and appropriate sewage disposal make diarrhoeal diseases one of the most important health problems in poor urban areas: the main cause of death of poor urban children under 5 years is often diarrhoeal dehydration. Malnutrition often occurs both as a cause and as an effect of diarrhoeal diseases. The need to buy food has been identified as one of the most important factors responsible for high rates of disease and mortality among the urban poor. Rural people may at least have a small plot of land on which to grow their own vegetables or to rear some animals, while the urban poor are dependent upon the cash economy. Infectious diseases such as measles and acute respiratory infections can have a significant impact on nutritional status, and the high level of indoor and outdoor crowding in poor urban communities means that transmission rates of infections are very high. For example, in one of the poorest slums of Port-au-Prince, the capital of Haiti, 40 per cent of children had already had measles by the age of twelve months. Often malnutrition in poor urban communities starts very early in life –

within the first six months. This is likely to be associated with early cessation of breast-feeding and the reliance on artificial feeding provided by surrogate mothers or siblings. The early introduction of bottle milk, encouraged by multinational companies' advertisements, and involving the use of contaminated water, is a particular threat to the urban infant. Data from the World Fertility Survey found that rural women breast-feed for two to six months longer than their urban counterparts. More substantial differences occur in Indonesia (nine months), Jamaica (seven months) and Thailand (eleven months). With poor urban women relying on work in the informal sector with no child-care facilities, this threat to infant health in cities seems likely to increase, given the economic decline in many, particularly sub-Saharan African, cities in the Third World.

Whereas the above health problems have been associated with the 'traditional' health profile of developing countries, certain health problems are now emerging within poor urban communities which have previously been considered typical of industrialized countries. An example of such health problems is mental ill-health. Mental ill-health is emerging as an important but neglected problem in developing countries. A recent analysis of the 'global burden of disease' is undertaken in the World Bank's (1993) *World Development Report*, which has the theme of 'investing in health'. The analysis uses 'disability-adjusted life years', which essentially take into account the expected duration of a condition, its severity and the age groups it is most likely to affect in order to produce a measure of the disability caused by morbidity (as opposed to years lost due to mortality, say). Mental illness has appeared as an important component of the burden of disease in all regions, particularly for young women.

Most of the community-based measures of the prevalence of mental ill-health in poor urban communities have been gathered in Brazil. A study of mothers with children under 5 years of age in the largest *favela* (shanty town) in Rio de Janeiro found that 36 per cent of mothers were probable cases of mental ill-health (Reichenheim and Harpham 1991). Several other studies have confirmed that low socio-economic status and gender (being female) are risk factors for mental ill-health in urban Brazil. There are several hypotheses regarding the causes of such poor mental health. Included among these are the erosion of the support of the extended family, the social dislocation caused by rural–urban and urban–urban migration, marital breakdown with resulting high numbers of female-headed households, women's 'double burden' of productive and reproductive roles, and environmental stress including high levels of violence. These social factors of urbanization in developing countries which are associated with mental health status are represented in the model in Figure 7.3. Clearly, health policy-makers, health planners and health providers have to take into account a very wide range of health problems when addressing the health needs of poor urban populations in the Third World.

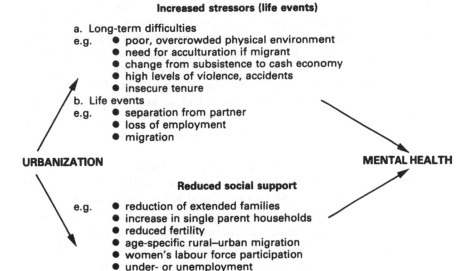

Figure 7.3 A model of social factors of urbanization associated with mental health in developing countries
Source: Harpham (forthcoming)

CURRENT DIRECTIONS IN URBAN HEALTH RESEARCH AND POLICY

Satterthwaite (1990) suggests that it is only our lack of knowledge about Third World cities which has produced so many generalizations about them. Certainly, the field of urban health research is relatively new, and there remain many gaps in our knowledge. For example, in terms of the physical environment and its effect upon health, there is a plethora of research on water supply and sanitation in relation to diarrhoea and intestinal worms, but there are very few studies on refuse disposal, drainage, crowding and their respective effects on health. Both road traffic accidents and domestic accidents remain under-researched. Turning to the health services environment, there are numerous case-studies which demonstrate patterns of utilization of health services in poor urban communities (including evidence to suggest that the traditional sector has adapted well to urbanization and remains an important source of health care – see Chapter 4). Little evidence is available on what poor urban residents pay for the wide range of health services (public, private, traditional, pharmacists) which they tend to use. Nor is there much information on the quality of urban health services (either professionally-defined or consumer-defined quality). A particular problem of city health services is the over-utilization of the out-patient departments of hospitals and associated under-utilization of primary health centres. One response to this has been the formation of a middle tier in the referral system: centres which are variously called 'reference urban health centres', 'filter clinics' or 'enhanced primary health care centres'. For example, in Cali, Colombia, such centres provide ambulatory surgery, normal deliveries, emergency care, rehabilitation, sterilization, neo-natal care and out-patient care. In Mexico City, similar centres were created out of necessity following the destruction of many hospitals by the earthquake in 1985. They have been successful and have therefore remained as key components in the overall plan of urban health facilities. Whether these centres promote a more 'rational' use of different levels of health service facilities needs to be evaluated.

Any urban health policy development has to address the respective roles of the Ministry of Health and the municipality. Often the municipality will come under the aegis of the Ministry of Local Affairs or Ministry of Urban Development or its equivalent. Thus, communications between the respective ministries are of paramount importance in the development of urban primary health care. An example of the problems due to lack of communication between these ministries comes from India. Several Indian cities including large centres such as Calcutta and Hyderabad have initiated slum improvement projects which consist of physical improvements to slum communities as well as community development and primary health care components. These projects are assisted by development aid from the United Kingdom and are often implemented by the municipalities, which, at the central level, come under the jurisdiction of the Ministry of Urban Development. The slum improvement projects have initiated innovative models of urban primary health care. For example, private practitioners who live in or near the slums, and who are regularly consulted by slum dwellers, are trained and paid an honorarium to run a 'primary health care centre' based in a community hall which serves as a nursery at other times. However, these developments have been made in the context of the slum improvement projects only and have not been accommodated into a comprehensive city health plan. The Indian Ministry of Health has had little input to the urban primary health care developments and thus has no vested interest in seeing them continue, or in learning lessons from them. This example is by no means unique. A typical area of conflict arises in health services *vis-à-vis* environmental health. Often different departments are responsible for these two areas of activity, and intersectoral co-ordination may be limited. The successful promotion of urban primary health care depends upon good leadership within the city, which can effectively overcome these constraints. In some cases this role can be successfully played by the mayor.

Municipalities in developing countries often have a very weak financial base. This problem is currently being addressed by the Urban Management Programme of the World Bank, the United Nations Development Programme and the United Nations Centre for Human Settlements, which is focusing on Jakarta (Indonesia), São Paulo (Brazil), Accra (Ghana) and other cities in developing countries. This programme includes the objective of strengthening the municipality in order that it can effectively promote and deliver urban primary health care. The role of municipalities in financing urban health services differs greatly between cities. In many African cities such as Dakar, Dar es Salaam and Lusaka, the federal government is involved in funding city health services. In some Asian cities such as Jakarta, Manila and Bombay, local governments are responsible for provision. In many Latin American cities such as Cali and São Paulo, there is a partnership between federal, state and local government. In Cali, for example, 5 per cent of the health budget (recurrent costs) comes from the national government, 25 per cent from the state government, 45 per cent from the local government and the remaining 25 per cent from user fees. This example is unusual in that a fairly high percentage of costs is covered by user fees. In other countries the percentage of costs covered by the controversial intro-duction of charges is well below 20 per cent, and there is growing evidence that utilization rates decline after the introduction of charges and that poorer populations in particular use services less.

The introduction of user charges in the health sector is often part of a 'structural adjustment' policy (see Chapter 3). Structural adjustment actually entails a package of numerous policies which typically encompass:

- devaluation of national currencies to discourage imports and encourage exports (to improve the balance of trade);
- reduction of government expenditure in the social sectors through privat-ization/user charges/withdrawal of subsidies;
- trade liberalization involving abolition of price and import controls and free entry for multinational corporations;
- privatization of government enterprises, 'retrenchment' (laying off) of workers, wage freezes and increases in prices.

Loans from the International Monetary Fund and the World Bank and aid from major Western donors are increasingly tied to the implementation of these policies; thus aid becomes 'conditional'. Even the World Bank acknowledges that structural adjustment is hitting poor urban populations the hardest (World Bank 1991). This is because: the urban poor are dependent upon their labour, rather than asset ownership, so are particularly affected by rising unemployment; the cuts in public expenditure (e.g. health), which are part of structural adjustment, tend to have a dispropor-tionate impact on the poor; and changes in tariffs and subsidies with resulting rising prices (particularly food prices) affect the urban poor, who are dependent upon the cash economy, more than their rural counterparts.

There is increasing evidence that, contrary to the World Bank predictions that the urban informal sector will thrive under free market conditions, the informal sector has in fact been squeezed from both the supply and the demand side. On the supply side, raw materials are scarce due to formal industrial producers now recycling waste products which were previously inputs of informal producers. On the demand side, the substantial contraction in demand among low-income wage earners who normally purchase goods and services from informal producers has had a particularly strong negative effect (based on a study in Zaria, Nigeria, by Meagher and Yunusa 1991).

Although there is currently no evidence of the impact of structural adjustment on the health status of low-income urban populations, it is likely that any increase in poverty will lead to declining health status. In order to measure such an impact, longitudinal, prospective studies are necessary, focusing on change over time among the same study populations. Such studies are notoriously difficult and expensive to implement. However, structural adjustment is a potential threat to the health of poor urban populations and poses a major constraint for future policies which attempt to improve health and health conditions in the cities of developing countries.

CONCLUSIONS

The rapid urbanization and growth of poor urban populations in developing countries are leading many national and international policy-makers to reconsider urban health planning. This shift was reflected by the fact that the World Health Organization chose the theme of urban health for its Technical Discussions in 1991. The WHO annual Technical Discussions bring together Ministers of Health from all countries of the world in order to discuss a topical subject of great importance. While in the 1980s it was necessary to look towards non-governmental organizations for innovative examples of urban primary health care, it is now possible to identify local and national governments that are addressing the problems of inequity and poor health in cities.

Research efforts over the last decade have illuminated more of the health problems of poor urban populations in developing countries. While much of the research focuses on the traditional areas of maternal and child health, relatively little evidence is available about the health problems of adolescents, adults and the growing number of elderly people in the cities. Gradually, more studies are available which demonstrate the differentials in health between different socio-economic groups within the city. However, many cities are still only able to produce aggregated data which shed very little light upon the range and diversity of health problems present. All health statistics must be examined critically in terms of the source of the data, the timing of the collection of data and the population which those data represent. In this current time of innovation in urban health, there is an urgent need for documentation of experiences and comparative evaluations of practices. The health problems of the poor have poverty as their root cause. Poverty, as expressed by lack of food, lack of finance, lack of education, poor sanitation and inadequate housing, is a major cause of the health problems described in this chapter. Changes in these basic causes of ill-health will do as much to improve health as health services will, if not more. However, even within the health sector many things can be done. Some have spoken of an 'inverse care law', whereby those in greatest need of medical care have the poorest access to it. Initiatives which successfully tackle this inequity should be supported and can be seen as a first step in improving health in the cities of the Third World.

REFERENCES

Harpham, T. (forthcoming) 'Urbanization and mental health in developing countries: a research role for social scientists, public health professionals and social psychiatrists'.

Harpham, T., Lusty, T. and Vaughan, P. (eds) (1988) *In the Shadow of the City: Community Health and the Urban Poor*, Oxford: Oxford University Press.

Lipton, M. (1976) *Why Poor People Stay Poor: Urban Bias in World Development*, London: Temple Smith.

Meagher, K. and Yunusa, M.B. (1991) *Limits to Labour Absorption: Conceptual and Historical Background to Adjustment in Nigeria's Urban Informal Economy*, Discussion Paper No. 28, Geneva: United Nations Research Institution for Social Development.

Reichenheim, M. and Harpham, T. (1991) 'Maternal mental health in a squatter settlement of Rio de Janeiro', *British Journal of Psychiatry* 159: 683–90.

Rossi-Espagnet, A., Goldstein, G.B. and Tabibzadeh, I. (1991) 'Urbanization and health in developing countries: a challenge for health for all', *World Health Statistics Quarterly* 44(4): 187–244.

Sattherthwaite, D. (1990) 'Urban and industrial environmental policy and management', in J.T. Winpenny (ed.) *Development Research: The Environmental Challenge*, London: Overseas Development Institute.

Tabibzadeh, I., Rossi-Espagnet, A. and Maxwell, R. (1989) *Spotlight on the Cities: Improving Urban Health in Developing Countries*, Geneva: WHO.

United Nations Department of International Economic and Social Affairs (1989) *Prospects for World Urbanization, 1988*, New York: UN.

United Nations Population Division (1984) *Urban and Rural Populations, 1984*, New York: UN.

World Bank (1991) *Urban Policy and Economic Development: An Agenda for the 1990s*, Washington, DC: World Bank.

—— (1993) *World Development Report 1993: Investing in Health*, New York: Oxford University Press.

World Health Organization (1991) *Health and the Cities: A Global Overview*, Technical Discussions on Strategies for Health for All, Background Document no. 2, Geneva: WHO.

8

THE HEALTH OF WOMEN
Beyond maternal and child health
Nancy Davis Lewis and Edith Kieffer

A NEW AGENDA FOR WOMEN'S HEALTH

Until quite recently, attention to women's health has most commonly been focused on the reproductive process and organs, and on the child-bearing period, which is generally defined as between the ages of 15 and 44. Occasionally, the health of girls has been considered, but again, primarily within the context of concern about assuring future reproductive health and the health of children. Most discussions of women's health, especially with respect to developing countries, have referred only briefly, if at all, to health problems apart from reproduction or to post-menopause. This continues to have important implications for health policy and planning, especially in an era of global economic recession, and at a time when comprehensive primary health care is being challenged and selective programmes are being offered as alternatives (Raikes 1989) (see also Chapter 9).

The reasons for this focus can be understood in relation to women's central role in bearing and raising children and, especially in developing countries, because of the tremendous vulnerability of women and children to morbidity and mortality during pregnancy, childbirth and infancy. It is estimated that more than 500,000 women die annually during pregnancy and childbirth, while many more are disabled. Specific initiatives designed to reduce maternal morbidity and mortality were established in the late 1980s by the World Health Organization (WHO) with support from international agencies and institutions. These initiatives and programmes fall under the general rubric of 'safe motherhood'. Activities have included increasing access to family planning, pre-natal and post-partum services; training of traditional birth attendants, nurse-midwives and other health workers to improve the safety of childbirth; immunizations and improved drug supply strategies; nutrition and health education campaigns.

These initiatives are crucial to improving the health of women, particularly in developing countries, where maternal morbidity and mortality constitute a substantial proportion of all causes of illness and death. However, they generally fail to consider non-reproductive aspects of women's health status, before, during, after or outside of the child-bearing experience. Nor do they consider the complex of social, economic and political issues involved in the way women carry out their productive and reproductive lives. Additionally, other socio-economic changes – such as the greater degree of education of girls and young women, the later age at marriage, and the wider acceptance and availability of modern contraception – also serve to reduce exposure to risks associated with reproduction. For women who do not bear children or have completed child-bearing, and in places where the proportion of illness and death due to reproduction is small, maternal definitions of women's health are inadequate at best. Even where reproductive mortality remains higher, maternal definitions are incomplete and need to be expanded as the mortality profiles of these regions undergo transition.

Concerns for maternal and child health are crucial. However, a significant and growing proportion of women's years and activities occur outside of the arena of reproduction. The focus of international concern must be expanded to include the quality of life during the entire life cycle within a context that addresses physical, mental, social and economic health. A broad definition of 'safe womanhood' must be adopted in which safe motherhood is central but not sufficient.

WOMEN, HEALTH AND DEVELOPMENT

The highly complex, reciprocal relationship between health and development has received considerable attention within the academic community and among development planners and practitioners. This relationship is explored more generally in Chapter 1 of this volume. Within societies, the development process affects individuals and groups differently, the commonly acknowledged example being the rich and the poor. Less often considered are the differential effects of development on women and men (Ferguson 1986). Furthermore, the contribution that men and women make to development differs overall, at various stages of the life cycle and also among socio-cultural groups. The value placed on these contributions, by society and by international aid agencies, also varies.

Women in development

Risks to women's health are linked to the quality of life and to the relationship of women to their economy, society and culture. Over the last two decades, the adverse effects of development on women, on the one hand, and the centrality of women's roles in economic development, on the other, have been acknowledged by activists, practitioners and, perhaps belatedly, scholars. A recently published anthology synthesizes much of the research on women and economic development carried out during these two decades. As the title, *Persistent Inequalities: Women and World Development* (Tinker 1990) suggests, global equity has not been achieved, and women continue to be discriminated against within the household, in labour markets and with respect to property rights.

While in recent years this central role of women in the development process has been acknowledged, operationalizing this awareness has been much more difficult. Baseline data on women's health and social risks and status, at all stages of the life cycle and in both the developed and the developing world, are often inadequate or missing. Such data, along with measures of gender differences in levels of health and accessibility and utilization of health care services, are essential for both the formulation of appropriate policy and the development of effective evaluation measures.

DEFINING AND MEASURING WOMEN'S HEALTH

The impressive compendium *The World's Women 1970–1990* (WHO 1991) compiled from various United Nations sources (e.g. the United Nations Children's Fund, the United Nations Fund for Population Activities and UNIFEM) includes a section on health and child-bearing and another on domestic violence. Data on health other than reproductive health are sketchy. Women outlive men everywhere except in Bangladesh, Bhutan, the Maldives and Nepal, and differentials in life expectancy generally increase with development. Given differences at age of marriage and differential life expectancy, an increasing number of women are widowed and dependent in later years. While the number of deaths due to infectious and parasitic diseases is decreasing, even in most parts of the developing world, maternal mortality remains high in many areas, and AIDS is an increasingly serious problem globally. In many parts of the world, women are malnourished at all stages of the life cycle as a result of both biological and cultural factors.

The statistics that are given in the above UN publication tell us very little about women's *total* health, even narrowly defined. The statistics are life expectancy, maternal mortality, infant mortality for both sexes combined, male and female child mortality, fertility, contraceptive use, trained attendants at birth and percentage of women who smoke. These data, as well as others sometimes available from other sources (such as birth intervals, proportion of women under 18 and over 35 bearing children, and minimum age of marriage), may be measures of the degree of control a woman has over her own life. Others, if available by sex (e.g. infant mortality, weight for age, life expectancy, immunization coverage, primary health care attendance, nutritional status and school enrolment), may be measures of gender inequity.

A male model

In both the biomedical and social arenas, questions concerning health and disease have been conceptualized on male terms or, at best, as gender-neutral. In the United States, for example, only recently has official and public attention led to an addressing of the discrepancy in health research funding and methodology that has focused almost exclusively on leading causes of morbidity and mortality predominating among males, e.g. cardiovascular disease, hypertension and lung cancer, rather than, say, breast cancer, a disease where there has been relatively little progress in terms of long-term survival in the past twenty-five years.

The majority of individuals in research populations and participating in clinical trials of new medications are men. Where females are included, the study results are often not analysed by gender. One response at the United States national level has been the recent formulation of the National Institutes of Health Office of Research on Women's Health. Not surprisingly, the bias is not only towards men but towards men in developed countries. Analysis of various studies in developed countries indicates that 45-year-old men who adopt a healthy lifestyle can extend their life by eleven years in comparison to men who do not. It is not known whether such behaviour would apply to women or to populations in developing countries. These findings thus apply to significantly fewer than 10 per cent of the global population. The reasons for this male bias and focus on the developed world, which are not limited to the United States, are myriad and related to the very fabric of society and its social, cultural and educational processes. Not unrelated is the dearth of female scientists and researchers, especially at the higher levels of the scientific establishment where research and funding decisions are made.

The issues that need to be addressed are legion. Researchers must employ a comprehensive definition of health and acknowledge magnitude differentials in the developed and developing worlds. Health status concerns include: sexually transmitted diseases, including AIDS; infertility and sub-fertility; safe contraception; gender differences in health throughout life, but particularly in adolescence and after 60; osteoporosis; cancer; cardiovascular disease; women and disability; substance abuse; mental health; accidents and injuries; suicide and para-suicide; occupational health; and nutrition and eating disorders. Related issues include: violence against women (domestic and community); prostitution; healthy environments at home, at work and in urban and rural settings; older women as care-givers; and abandoned older women. Issues of more limited geographic scope, but no less important, are those that face women in regions characterized by extreme social and political disruption – e.g. war, and refugee and migrant women – and, for populations in Africa and some parts of Asia, female circumcision and mutilation. Health service concerns include: gender differences in access to and utilization of health care; ability to pay for care; health insurance; and women's health knowledge and practice, ranging from perception and use of traditional medicine to involvement in Western biomedical systems and access to counselling.

Women or gender?

There are differences between gender studies and the study of women. Simplified, the former concern themselves with comparison and analysis of factors related to gender, that is, between males and females. Gender is a socially constructed category which may vary from society to society (sex, on the other hand, is biologically determined). In the health arena, gender studies might include studies of infant mortality rates of girls and boys in the same population. Studies of gender and health, especially those that go beyond obvious differences and look for causal factors, have much to reveal about the differences between men and women on a number of dimensions including health status, health risks and health care delivery. They will also highlight extant inequities between women and men. They may suggest factors responsible for the differences or provide hints at aetiology, as we search to explain complex patterns of health and ill-health.

The study of women's health (sometimes carried out explicitly to make them visible), on the other hand, focuses on specific women's health issues, risks, or the use of health care in a study community or comparatively across communities. Both approaches – gender studies and the study of women – have a place in a 'new geography of women's health', although one or the other may be more appropriate at different scales of investigation or with respect to particular topics. It might be appropriate, on the one hand, to explore the differential access to health care of men and women in a Third World city, and, on the other, to study the use of traditional birth attendants by women in a rural area of a given nation. One could study gender differences in diseases of later life such as cardio-vascular disease or cancer at the national level, or study women's experience of menopause in several developing countries. In both geography and health, women have until recently been largely invisible when it has come to developing theory and defining primary research questions. Our gender blinders must be removed, questions redefined and theory re-examined.

This is not the place for a thorough review of the debates central to the broader arenas of both geography and gender or feminist geography, but it is our premise that the issues raised are so critical that gender awareness not only must be incorporated into conventional methodological approaches within medical geography, but also must help to redefine its questions. It is time that feminist geographers move beyond considerations of welfare geography and directly address questions of women's health. As we propose in the following sections of this chapter, one way in which medical geographers can contribute to a greater understanding of women's health is by examining health in terms of the spatial and temporal spheres in which women interact. The small but growing body of literature within geography on redefining women's space needs to be explored in this context. Approaching the topic from both fronts promises to advance an exciting new frontier and one that can be informed by other feminist scholarship on health and health-related topics.

GEOGRAPHY OF WOMEN'S HEALTH

To date, the geographic contribution to research on women's health has been curiously sparse, even in comparison to other social sciences. What little has been done by medical geographers has focused on the reproductive period, infant mortality and child survival. Some work has addressed women as care-givers, and in some studies on health care and health care utilization, men and women have been differentiated.

In many disciplines, the most common way of investigating women's health has been defining it in terms of the life cycle. This framework has been criticized for its biological reductionism, as it tends to focus on the reproductive process and also tends to medicalize women's lives and health. The richness of gender differences, class, geography, religion, politics and sexual preference may also be lost (Clarke 1990). While there are limitations to the life-cyle approach, Lane and Meleis (1991) illustrated its contribution to examining women's health issues in developing countries. The authors assessed the health risks, perceptions and resources available to women in their specific, and often culturally defined, societal work and family roles and responsibilities, from girlhood to womanhood to old age. If a life-cycle or life-phase approach is undertaken, care must be taken to consider socio-cultural definitions of life stages. In many developing countries, women's roles and activities transcend stages of reproductive development. Taking this fact into account may contribute to moving beyond reproductive definitions of women's health.

Important contributions that geographers can make to an expanded definition of safe womanhood can be found in the classic geographic areas of investigation – space and time. As we shall suggest, however, delimiting the spatial and temporal spheres in which women interact on a daily basis soon becomes highly complex, in large part because of their multiple roles and responsibilities at any one time. Certainly, the typical delineation of 'home' and 'work' is inadequate in both rural and urban regions of the developing world. Even trying to define 'home' is problematic. Does it include only what happens inside the physical structure – inside the door? (Much social science research stops outside the door.) Or does it include the compound and the settlement or subsistence production area? As we shall discuss later, much informal and some formal production activity occurs within the home, and much social reproduction occurs outside of it.

We must ask what women (and men) *do*, as well as where, when and why. We must then assess how these activities expose individuals to various health risks, remembering that neither women nor men are homogeneous groups, and that age, educational level, ethnicity and a host of other factors must also be considered. Furthermore, we must ask what resources exist to promote health and to treat ill-health, how they are controlled, and how decisions are made concerning their use and who benefits.

While not suggesting that a political–economic framework is the only one to employ in analysing women's health (although it may be a useful one), nor adopting it explicitly here, using its categories may allow us to begin to develop a time/space framework for exploring risks to women's health. Women's health may be conceptualized in terms of their role in biological reproduction; in production in agriculture and the formal and informal sectors; and also in social reproduction, which loosely defined is activity related to the survival and functioning of the household (Raikes 1989).

From a geographic perspective, we can think about risks to a woman's health (and to a man's health if we are engaged in gender studies) in terms of her movement through space as she performs her varied functions. Similar models that investigate spatial and temporal aspects of risk have been proposed by medical geographers. The most relevant here may be the *human behaviour/disease ecology* model proposed by Robert Roundy (1987) for mobile populations of north-west Ethiopia. Roundy looked at risks for adult males, adult females, working children and non-working

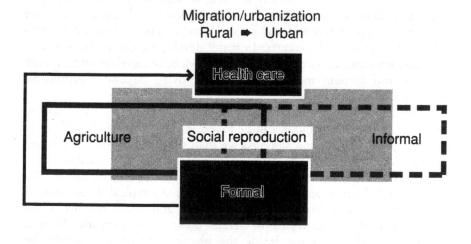

Migration/urbanization
Rural ➡ Urban

Ageing populations

Figure 8.1 Women's spheres of activity

128

children in terms of time spent in cells/spheres identified as individual, household, compound, settlement, production area and further-ranging contact. At the individual level, a 'self-specific environments' model might also be adapted (Armstrong 1978).

As noted, in addition to space, any model must also include considerations of how women allocate their time. Little is known about how women balance their various biological, economic and social roles, how decisions are made and how this affects health. Time allocation studies on market production and home production which have been carried out in the developing world suggest that women spend more time in total production than men. These are primarily rural studies; little is known about urban/ rural differences. While women spend somewhat less time in paid productive activities, they spend considerably more time in home production. Because of a 'double work' day, the difference can sometimes be considerable. A review of studies in Bangladesh, Java, Nepal, the Philippines, Botswana, the Central African Republic, Ivory Coast and Tanzania has indicated that in all cases women worked longer than their male counterparts.

Health risks and the location and types of women's activities

Identifying health risks in the places where women spend their time and assessing the degree of risk based on temporal considerations address key geographic questions. For women, the allocation of time changes much more greatly over the life cycle than it does for men (Leslie et al. 1988). Another temporal consideration in looking at women's health must be the seasonal nature of much agricultural activity, their extra burdens at harvest, and pre-harvest hunger which compounds anaemia and other conditions and may influence pregnancy outcomes.

The spheres of activity of women during their daily lives influence the nature and degree of their exposure to a variety of health risks (Figure 8.1). Many studies of household and occupational health risk have been based on an assumption of two-parent families with men working outside of the home and women working inside of the home. This model, which has always been of doubtful validity in many populations, is increasingly irrelevant in both developed countries (recent estimates note that only 7 per cent of the United States population live within such a household) and developing countries, where women's work in the agricultural and informal sectors is often not considered (Lane and Meleis 1991). A WHO survey reports that 22 per cent of households in Africa, 20 per cent of households in the Caribbean, 18 per cent of households in Asia, 16 per cent of households in the Middle East and 15 per cent of households in Latin America are headed by women. A side-effect of economic development schemes has been to increase male wage labour and often migration, drawing men away from traditional field labour, thereby increasing the overall workload of women as they are drawn into the fields, for production of food for trade and domestic consumption. Raikes (1989) noted that the research into female-headed households and their impoverishment in the developing world has not been accompanied by work on their comparative health status.

Studies of health risk are generally most easily conducted in the public arena, where male activities tend to dominate. In both developed and developing countries, to the extent that women's social, economic and domestic activities occur within the home, they are therefore less likely to be studied or represented statistically. Women generally account for almost all child care and household labour, and in developing countries often haul most or all of the water and household fuel, raise much of the food for domestic consumption and, additionally, often assist men in field labour and engage in trade and marketing activities. Since our investigation cannot stop at the doorstep, and since a typology that would look at risks to women's health at home vs at work would not incorporate the complexities typical of many women's lives, it is useful to try to conceptualize risk in terms of the *environments* associated with the spheres of social reproduction and of production in agriculture and the informal and formal sectors. As we are attempting to address women's health beyond maternal and child health, we do not consider reproductive function, but acknowledge that this is a false distinction. Many women are either pregnant or breast-feeding as they carry out their productive and socially reproductive functions, during a large portion of their lives.

While it is tempting to attempt to build a conceptual model that clearly demarcates the activity spheres of women's lives, the spatial and temporal complexity of these spheres and the multiple roles of women quickly frustrate the attempt. It should be noted at the outset that, in contrast to a number of health risks that are specifically related to gender differences in social status, gender differences in labour may have less to do with differences in social status of men and women than with culturally based divisions of labour between men and women. Specific labour tasks lead to specific environmental risks and thus to different causes of morbidity and mortality. It is important therefore to understand the socio-cultural and environmental context underlying men's and women's life circumstances in order to identify the likely health risks they face.

In both industrialized and developing countries, the percentage of women in the formal labour force is growing, while their home-based work and other activities continue. In addition to the specific health risks within each activity sphere, women appear to be faced with a double or triple burden of risk as they fulfil multiple labour roles that combine with each other the risks of each sphere alone, as well as enduring the physiological consequences of child-bearing.

A multitude of questions could be posed concerning women's health risks. Here, based on a relatively sparse literature, we will suggest a few questions that need to be asked with respect to activity spheres associated with agricultural, informal and formal production and social reproduction.

Production

In general, the health of workers in developing countries has received little consideration, regardless of gender, and this is particularly true for workers in the informal sector. In the formal sector, occupational health and safety standards are weak, if they exist at all, and work-related illness and injuries receive little attention.

Agricultural production

Agricultural production includes subsistence production, which for women overlaps with social reproduction, various forms of cash cropping, plantation agriculture, and wage labour on commercial estates. It may also include livestock rearing. By its nature, agricultural production is primarily a rural activity, although where land exists, urban gardening can provide an important source of nutrients in the squatter settlements of Third World cities. The risks to women's health with respect to agricultural production are varied and in many cases are similar to those of their male counterparts, although the duration of exposure and its spatial and temporal pattern may vary. These risks include exposure to pathogens in the environment, to extreme temperatures and to insolation, the carrying of heavy burdens and exposure to toxic agricultural chemicals. Due to social status and cultural differences, women may have to travel further to fields and may be carrying infants or toddlers with them. In some instances, technological advances have actually increased women's work by increasing the area under cultivation. However, women often have unequal access to training, credit and the benefits of technology.

The increase in cash cropping in many parts of the developing world has meant that there has been a loss of land under subsistence production, a move to more marginal land and an increase in distance from home to subsistence field. Ferguson (1986) has pointed out that this marginalization has had a negative impact on women's health in Kenya, also noting that compulsory education for children has increased the workload of women. Raikes (1989) outlined the impact on women's health of plantation schemes, which have had negative effects both on the women left as head of households and on those who accompany their husbands to the plantations. These effects include little or no access to health care, no land for subsistence crops (which leaves no alternative to buying expensive food in the company store), sexually transmitted diseases and domestic violence. The more general socio-economic impacts of changing agricultural systems and other economic activities such as mining, which may increase male migration and impoverish women, must also be part of the analysis. The burden and risk to the household of ill and disabled labourers returning from mines and agricultural schemes may also be significant.

Informal production

Informal-sector activity is that which occurs outside of structured employment, and it is usually outside government control and taxation. It is often intermittent and irregular. While both men and women are engaged in informal activities, particularly in the cities of the Third World, women outnumber men in these activities. For women much of this activity actually occurs within the home and includes child care, sewing, washing and other similar activities, and the production of craft or food items that are then typically sold on the street. It may also include activities such as street vending of other items, domestic labour in the homes of others, alcohol brewing, waste hauling and prostitution. While relatively few studies address women in the informal sector, it appears that there are blurred boundaries between unpaid domestic labour (or social reproduction) and informal activities. Women sell in the informal market-place the skills that they normally practise at home. They are providers of food, beer, child care and sex-companionship in both arenas (Nelson 1979).

In one sense, the home has been restructured as a workspace, and the informal activities take place in the same place and at the same time as social reproduction. Risks include burns from cooking fires, indoor air pollution, poor ventilation, poor lighting, possibly hazardous materials in home manufacture, and exposure to the infectious diseases of childhood. Street vending has its own set of risks, including traffic, pollution and exposure to violence. Prostitution brings with it risks of physical and emotional violence, sexually transmitted diseases and now HIV.

Sexually transmitted diseases have always been a risk for women engaged in prostitution. With the advent of HIV, the problem has assumed new proportions. The statistics for HIV sero-positivity among female commercial sex workers in a number of developing countries are staggering: up to 80 per cent in Kenya, 32 per cent in Bombay and up to 44 per cent in various provinces in Thailand (cited in De Bruyn 1992) (see also Chapter 5). But the continued focus on these high rates in prostitutes and their use in well-intentioned public health campaigns may continue to stigmatize women as transmitters of the disease.

De Bruyn (1992) presented a comprehensive picture of the issues surrounding women and AIDS in developing countries. She noted that the stereotypes related to the disease mean that women are either blamed for the spread of the disease or not recognized as potential patients. She also noted that the psychological and social burdens are often greater for women than for men, largely related to their roles as mothers. Furthermore, women's commonly low socio-economic status and lack of power make it difficult for them to use prevention measures. It is important to note here that engaging in sexual activity for money or other compensation in the Third World is not limited exclusively to a sub-set of commercial sex workers. Poor women may not only be at risk from non-monogamous partners, but because the only recourse that they have when an urgent need for cash arises may be to engage in sexual activity. It may also be an additional 'service' expected as part of badly needed employment. Women may also be at increased risk of HIV because they are the primary providers of care, such as traditional birth attendants. They also bear disproportionately the social, economic and health-risk burdens when family members are infected.

Formal production

As with the informal sector, as women enter the formal workforce it is unlikely that their health risks mirror those of men, since they often fill different niches in the labour force. While formal-sector activity is predominantly male, even in Latin America, Browner (1989) noted that 25 per cent of the paid workforce is female. Within the same occupations, social, biological and coexisting factors in the home environment may make women differentially vulnerable to morbidity and mortality. As noted in Chapter 1, the rapid pace of socio-economic change in many developing countries has occurred without a concomitant development of occupational and environmental health-and-safety regulations or health care delivery and financing mechanisms. While this problem affects both men and women, low levels of women's education and social status make it even less likely that women will actively seek changes in unhealthy environments.

Little information exists about health risks to employed women in the developing world. In Swaziland, for example, women working at a multinational-owned pineapple cannery, and most of whom were single mothers with a number of children, received very low wages and were given little or no protective clothing, and only some of them were provided with housing. There was no medical data available on these women. In many countries, women are often hired at lower wages then men. Leslie *et al.* (1988) suggested that, because of their multiple responsibilities, women sometimes choose lower-level jobs because of their need for time flexibility. Women are also subject to a variety of occupational exposures, some of them the result of the export of hazardous industries from the developed countries. They are also subject to abuse from co-workers and to societal violence, especially in cities. Traffic accidents, for both pedestrians and passengers, and perhaps especially the risks of travel by motorcycle or motorbike, should be added to occupational exposures, although women may be less at risk than men because of less ready access to a vehicle.

Social reproduction

Social reproduction, the activities related to the functioning of the family and household, takes place both within and outside of the home. Not surprisingly, risks associated with social reproduction include many that have been considered the 'traditional' health risks of women, such as burns from cooking and heating, especially since the introduction of fossil fuels; indoor air pollution; risks associated with the processing and storage of foods; exposure to crowded, poorly ventilated and unsanitary conditions; and childhood diseases and respiratory infections. Outside of the house or compound these risks related to social reproduction may be associated with everything from gathering water – and, for instance, being exposed to the trematode which causes schistosomiasis – to obtaining health care. Ferguson (1986) noted that in a district of rural Kenya 70 per cent of the water collected was collected by women, and that meant carrying 20–25 kg of water 3.5 km over rough terrain three times every two days. While women carried water on their backs, men were much more likely to carry it on a bicycle, cart or donkey. Women also often put the health and nutrition of their families first, suffer nutritionally and are the last to seek medical care.

Domestic violence occurs within the social reproductive sphere and it is a serious problem in all parts of the world. Female circumcision also occurs within this arena, most often perpetuated by women fearful for their daughters' future marriageability, as does the socialization of young women into the subservient ways of their mothers.

Urban areas

Urbanization, which is occurring at a rapid pace in much of the Third World, may have different implications in terms of health risks for women and men (see Chapter 7). Cities of the developing world are growing, often much more rapidly than the nations of which they are a part, due both to rural–urban migration and to the growth of urban populations themselves. A consideration mentioned previously was the impact of male wage migration to the cities, leaving female-headed households in rural areas. Workloads for rural women have increased, exacerbating health and nutritional problems for them and their children.

Migration is not solely a male phenomenon and, historically, it has frequently involved females, as it does today. A move to the city, commonly to a squatter settlement, may expand the activity space for both men and women and make it more complex, but not necessarily on the same dimensions. Women, because of their multiple responsibilities and concentration in the informal sector, may find themselves travelling great distances by poor public transport or on foot for many reasons including accessing health care for their dependants.

Crowding in squatter settlements may exacerbate risks from poor access to water and sanitation, stresses of urban life, domestic violence in the home and social violence and crime in the neighbourhood. In addition, the traditional supports of family and social networks may be missing, and the traditional formal institutions and infrastructure of the city or state may also be absent. The issues of housing and habitat may be even more critical for women than for men in Third World cities because of their lower status and because more of their productive and socially reproductive activities take place within the home. Little attention has been paid to the effects of such stresses on the mental health of women, nor to the growing number of elderly women, some of whom are forced from rural areas due to divorce or to the death of their husbands and live without resources or family. For biological and social reasons, the ageing populations of the Third World cities are predominantly female.

Health care services

The crucial role of women in providing for overall family and community health has often been acknowledged. It is concomitant with women's socially reproductive function. Seventy-five per cent of all health care occurs at the family or individual level (Leslie *et al.* 1988). Historically, women have had a central role to play in all aspects of health as wives, mothers, grandmothers and traditional practitioners. Repeatedly correlations have been found between maternal education and child survival, although it may be questioned whether this is due primarily to education itself or to other socio-cultural factors. Furthermore, the proportion of women who are health practitioners, and their roles within the health care hierarchy, may be a measure of gender equity, or lack thereof, in different societies. There remain, however, many questions with regard to women's participation in and use of health care services that lend themselves to geographic investigation.

A significant portion of medical geographic scholarship has focused on accessibility and utilization of health care services, in both the developed and the developing world. Utilization studies have sometimes, but not always, included gender analysis, at least to the point of differentiating utilization by males and females. Less focus has been placed on differences in accessibility – geographic, economic and socio-cultural – for men and women. In addition to the activity spheres identified for production and social reproduction, a 'health care services' sphere of activity is proposed to help elucidate the complex factors involved in women's health and health care behaviour. Here the focus would not be on health risks but rather on questions related to the absolute and relative location of facilities, hours of operation, cost, transportation and cultural sensitivity, as well as differential use of various forms of services – for example, Western medicine in comparison to traditional practitioners.

Specific services for women in developing countries have been designed primarily as services for maternal and child health. Where resources are scarce, the health of the mothers often receives considerably less attention than the health of their children. Illness not typically part of maternal and child health (MCH) falls through the cracks. For example, women are at highest risk of cervical cancer after menopause, but these are the women who are no longer being seen at family-planning or MCH clinics.

Women use health care services themselves, and as part of their socially reproductive function they seek health care for their children, the elderly and other family members. What are the dimensions of this? Are all visits to a health care facility primarily to receive care? Visits to a clinic may have a social function as well. Do men and women, boys and girls have the same access to care? Does this depend on distance, cost, illness or some combination of these factors? Within the household, who controls the resources for health care?

Many studies in developing countries have noted that the medical traditions of men and women may be very different. Lane and Meleis (1991) found in research in rural Egypt that lower-status individuals, who often tended to be young females, received only home remedies. A higher-status individual, males of any age or an adult mother of sons, would be taken to a Western-style medical practitioner even at some distance from the village. Young (1989) noted that on Goodenough Island in Papua New Guinea boys were twice as likely to be taken to out-patient facilities as girls. The islanders explained this by the fact that boys have higher social value. Men are the owners of the village, its inheritors, the 'house-posts', whereas women are 'bouncing coconuts' who leave their home villages to live in the villages of their husbands. De Bruyn (1992) reported that, while the incidence rates of HIV in Africa are similar among men and women, almost all the AIDS hospital beds are occupied by men.

CONCLUSION

Women's multiple roles and responsibilities expose them to myriad health risks that are poorly understood. At the global and national scales, for the developing world, mortality data are often not available by sex (and often not by age). Morbidity data, poor as they often are, are even less available by sex. We must begin to develop ways to accumulate good data, in order to identify the complex patterns of health, female and male. At regional and local scales, other kinds of investigation, ideally utilizing both biomedical and social science approaches in cross-disciplinary research, will help us to understand these patterns. The collection and analysis of data on women will also provide visibility which can ultimately be translated into empowerment.

We do not advocate an approach to research, policy or planning that focuses solely on women. It has been argued, in fact, that much of what has come under the rubric of maternal and child health and 'safe motherhood' has been viewed as a selective primary health care 'magic bullet', and that it has ignored the needs of men. We would argue, however, that this same approach has also ignored many of the health needs of women. We urgently need to understand patterns of health and disease of both men and women throughout life. We must address the invisibility of women, especially beyond their reproductive health, and encourage the development of policies that will lead to a comprehensive approach to the world's most pressing health problems and needs. As economic and environmental issues are linked, so safe and healthy women contribute to economic growth and environmental protection. Only with appropriate attention to the health of women can they participate in national development to their fullest potential.

REFERENCES

Armstrong, R.W., Kanna, K.M. and Armstrong, M.J. (1978) 'Self-specific environments associated with nasopharyngeal carcinoma in Selangor, Malaysia, *Social Science and Medicine* 12D: 149–56.

Browner, C.H. (1989) 'Women, household and health in Latin America', *Social Science and Medicine* 28(5): 461–73.

Clarke, A.E. (1990) 'Women's health: life-cycle issues', in R.D. Apple (ed.) *Women, Health and Medicine in America: A Historical Handbook*, New York and London: Garland, pp. 3–39.

De Bruyn, M. (1992) 'Women and AIDS in developing countries', *Social Science and Medicine* 34(3): 249–62.

Ferguson, A. (1986) 'Women's health in a marginal area of Kenya', *Social Science and Medicine* 23(1): 17–29.

Lane, S. and Meleis, A. (1991) 'Roles, work, health perceptions and health resources of women: a study in an Egyptian delta hamlet', *Social Science and Medicine* 33(10): 1197–208.

Leslie, J., Lycette, M. and Buvinic, M. (1988) 'Weathering economic crises: the crucial role of women in health', in D. Bell and M. Reich (eds) *Health, Nutrition and Economic Crises: Approaches to Policy in the Third World*, Dover, MA: Auburn House, pp. 307–48.

Nelson, N. (1979) 'How women get by: the sexual division of labour in the informal sector of a Nairobi squatter settlement', in R. Bromley and C. Gerry (eds) *Casual Work and Poverty in Third World Cities*, Chichester: Wiley.

Raikes, A. (1989) 'Women's health in East Africa', *Social Science and Medicine* 28(5): 447–59.

Roundy, R. (1987) 'Human behavior and disease hazards in Ethiopia: spatial perspectives', in R. Akhtar (ed.) *Health and Disease in Tropical Africa*, London: Harwood, pp. 261–78.

Tinker, I. (ed.) (1990) *Persistent Inequalities: Women and World Development*, New York and Oxford: Oxford University Press.

World Health Organization (1991) *The World's Women 1970–1990: Trends and Statistics*, New York: United Nations.

Young, A. (1989) 'Illness and ideology: aspects of health care on Goodenough Island', in S. Frankel and G. Lewis (eds) *A Continuing Trial of Treatment: Medical Pluralism in Papua New Guinea*, Dordrecht, Boston, MA, and London: Kluwer, pp. 115–39.

9

MATERNAL AND CHILD HEALTH CARE STRATEGIES

Penny Price

Every day almost 40,000 children die. Such deaths are overwhelmingly concentrated in the developing world. In 1990 mortality rates for the under-5s were 284 and 180 per 1,000 live births for Mali and Bangladesh respectively, a stark contrast to the 9 per 1,000 of the United Kingdom. Every day over 1,000 women die from problems related to child-bearing. The disparity between developed and developing countries is even more striking. India and Bangladesh have more than 100 maternal deaths per 100,000 women per year, compared to just 3 in the United States. Many millions more women and children suffer untold temporary and permanent disability and ill-health. The vast majority of this death and suffering is preventable.

Child ill-health, described as 'the silent emergency', has been a focus of attention now for several decades, unlike 'the neglected tragedy' of maternal health, which has only recently received due recognition. In the late 1970s and early 1980s short-term interventionist strategies, the Child Survival Initiative (CSI), appeared to show encouraging improvements in child health. However, recession and the debt crisis served to undermine such efforts in the latter half of the decade, and similarly hindered the development of the maternal programme, the Safe Motherhood Initiative (SMI). The persistence of high rates of maternal and child mortality reflects the need to question the value and long-term effectiveness of these strategies. Concern over the acceptability, accessibility and uptake of programmes is central to any discussion on the nature and cost-effectiveness of such a high-level response. However, if targeted universal coverage is to be achieved, then the causes of the problems need to be fully understood. Just why are mothers and children so vulnerable?

THE CAUSES

Mothers and children have long been recognized as particularly vulnerable sections of the population deserving special attention. Priority has been given to them as they represent the least powerful members of society. For

138

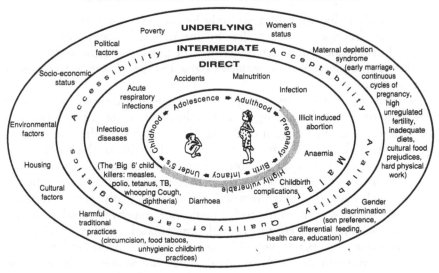

Figure 9.1 The causes of maternal and child mortality and morbidity

biological reasons the burden of disease and death seems to fall upon them, and all too often their needs have been neglected or misunderstood. Promotion of their health is important not only as an investment in the social and economic development of a community but also for the future of a nation as a whole.

The reasons behind such vulnerability are multi-layered and reflect the interwoven nature of the life paths of both mother and child (Figure 9.1). The direct medical causes are well known. All are largely preventable. Individually, each of the conditions is serious in its own right; however, it is worth emphasizing the synergistic relationship of infective, nutritional and parasitic diseases, as all act together to reinforce each other in a vicious infection–malnutrition cycle.

To date, attention has focused on the clinical causes, and yet this is only a partial picture. Behind the medical causes are the logistic causes – failures in the health care system, i.e. the availability, accessibility and quality of appropriate preventive and curative services – and beyond these are the underlying political, socio-economic and cultural factors that together determine health-seeking behaviour. Yet the most fundamental causes of all are poverty and underdevelopment. If the long-term situation of maternal and child health is to be improved, then these deep-rooted causes need to be addressed.

The health and life path of the young child are inextricably bound up with those of the mother, as is evident in the very label 'MCH' (maternal and child health). The health of the mother is not only important for its own sake but crucial for the well-being of her children. The health of both is profoundly influenced not only by their biological inheritance and individual circumstances, but also by the physical, socio-economic and cultural environment in which they live. One powerful factor underlies and reinforces all of these, namely women's status and the inherent discrimination against them.

Women are disadvantaged from birth through to adulthood, where their contributions and roles are largely undervalued. Yet women play a vital role in society, being the linchpin of the family, coping with the dual roles of production and reproduction. Recognition of the value of these productive roles (as farmers, health carers, housekeepers and educators, as well as paid workers) has been slow in coming – unlike recognition of their reproductive roles, which have long been 'revered'. Fertility is greatly prized. In many societies it is the woman's only path to influence, improved social status and personal achievement. However, high unregulated fertility is the very condition that perpetuates women's poor status and dissatisfaction.

Gender discrimination and a woman's status and health are all intricately entwined, and together significantly influence the health and survival chances of the child. The dice are heavily loaded against females as social customs buried deep in tradition clearly illustrate a preference for male children, through differential workloads, feeding, education and even health care. Social programming from birth conditions girls into a role of sub-servience. In childhood, various 'rites of passage' subject them to the lifelong physical and psychological damage of circumcision. Early marriage, continuous cycles of pregnancy, inadequate diets exacerbated by cultural food prejudices, infection and hard physical work leave women in a constant state of 'maternal nutritional depletion'. With these effects reinforced by low social status, discriminatory traditional practices and poor environmental conditions, most women are well on their way down what has been termed 'the road to maternal death'.

The considerable spatial and temporal disparities in maternal and child health that result from such a complexity of factors clearly demonstrate the importance of targeting those at most risk. Tackling the fundamental causes would be the long-term solution; however, a more immediate response lies in short-term interventions that now have the potential to prevent and reduce the incidence of childhood disease and complications during pregnancy and childbirth.

THE RESPONSE: THE DEVELOPMENT OF MCH STRATEGIES

It was only in the 1970s – which saw an increased awareness of the importance of health and development, and marked the beginning of a new era in health – that mothers and children became a focus of attention. A realization that the wholesale transfer of Western-style curative health care was failing to meet the needs of the majority of people in developing countries led to the introduction of the primary health care (PHC) concept, with its accompanying shift of emphasis from hospital-doctor-oriented care to community-based primary health care. The World Health Organization (WHO) and United Nations Children's Fund (UNICEF) conference on PHC declared that maternal and child health care was to be one of the essential components of PHC. It would be the key to achieving an acceptable level of health throughout the world, in the comprehensive strategy to achieve 'health for all by the year 2000'. The subsequent trend has been to incorporate the development of MCH care in the expansion of PHC at the local community level, with adequate provision for referral to the tertiary level, thereby providing both generalized and specialist care. Consequently PHC, by emphasizing equity in health and universal access for women and children to health care, has become the guiding philosophy for health care development in the 1980s and 1990s: the health dimension of 'real development'. With over half the potential users of PHC system being mothers and children, its importance for the future population of the developing world is clear.

Following the 1978 Alma Ata declaration of PHC, UNICEF, with its mandate to reduce infant and child mortality, encouraged the uptake of a 'targeted PHC strategy' – using specific measures to tackle specific issues within PHC. They emphasized the promotion and use of low-cost measures to benefit those most disadvantaged, mothers and children, and subsequently launched a 'child survival and development revolution' (CSDR). The major concept of the CSDR is that people can change their own lives through increased awareness. The success and effectiveness of programmes depend upon information and know-how being freely available, at the disposal of the majority. Thus very considerable improvement in the well-being of the world's children could be achieved within a relatively short time, using relatively simple and inexpensive techniques, particularly pertinent at a time of economic recession.

These 'cutting-edge' techniques are widely known under the acronym of GOBI–FFF. Initially these were just four simple measures concentrating on child health, namely Growth monitoring, Oral rehydration therapy, Breast-feeding and Immunization. However, with the increased recognition of the role of mothers in promoting their own and their children's health, GOBI was subsequently broadened in scope to include the equally vital but more difficult and costly elements of Food supplements, Family spacing and Female education.

The four principal GOBI strategies are distinguished by being affordable, available, almost universally relevant and able to achieve rapid results. Yet their most notable contribution towards a child health revolution is through their combined impact, considerably greater than the sum of their individual contributions. A concerted attack by the GOBI strategies could potentially break the self-perpetuating synergistic alliance of malnutrition and infection. A breast-fed and immunized child is less likely to contract diarrhoeal infections; a child with fewer episodes of diarrhoea is less likely to be malnourished; and a child who is less malnourished is less likely to succumb to infections.

By the mid-1980s global optimism strongly endorsed the feasibility and effectiveness of GOBI–FFF. UNICEF estimated that 4 million children had been saved through such preventive measures. However, it acknowledged that, if nationwide programme coverage was to be achieved, increasing coverage levels from an existing 20–30 per cent to 80 per cent targets by 1990, then progress would need to be accelerated, and the concept of 'going for scale' was introduced. Political and social breakthroughs were advocated as essential to the spread of GOBI and for the full potential of CSDR to be realized. These breakthroughs focused on three integral elements: the need for greater serious political and professional commitment, with GOBI–FFF programmes endorsed as national economic and political priorities; the global mobilization of people at all levels of society through community participation; and the improvement of communication channels. However, popular demand and political will alone are insufficient. The effective implementation of such strategies requires increased levels of awareness and the commitment of resources. In the absence of any concerted integrated approach, further improvement in the general levels of maternal and child health is difficult to envisage.

Increasingly the effectiveness of these interventionist solutions is being called into question. If theoretical coverage targets are to be achieved, then it is pertinent to examine what people are actually doing in reality, their health-seeking behaviour and the factors affecting the use of the strategies, or more significantly the barriers to their non-use. The obstacles to availability, acceptability and accessibility are crucial factors in the uptake of MCH strategies, as the reaction people have to health services in general remains true for specific services such as immunization and family planning.

The availability of a specific strategy and its initial coverage may be good; but, to reach those identified as at most risk and therefore in most need, an intervention must be acceptable. MCH strategies have been largely criticized for their inappropriateness. A predominance of male health professionals may not be fully aware of the needs of mothers and children, and may not be acceptable for women, particularly in Muslim areas. Attitudes and beliefs are often seen as the greatest obstacles to acceptance. The indiscriminate promotion of family planning has in the past aroused much suspicion. Considerable debate has also centred on the ethics of 'social marketing' campaigns that impose Eurocentric, 'foreign' ideas and means upon people. Likewise the widespread use of readily accessible indigenous medicine can restrict the uptake of such Western, allopathic services as immunization.

Accessibility and affordability are crucial concerns. Distance can discourage people from even trying to reach care. Rain, remoteness and harsh terrain produce extra transport difficulties, especially pertinent for a pregnant woman accompanied by a sick infant, and life-threatening for a woman experiencing childbirth complications. Increased distances frequently imply increased total costs. The total actual cost of obtaining preventive care involves not just the costs of care – treatment, drugs and supplies – but also the opportunity costs of time and transport, especially if referral is necessary. Consequently, services supposedly 'free of charge' are in fact expensive. Underutilization may also be due to *perceived* quality of care. Poor-quality services – suffering from shortages of suitable staff and inadequate supplies and equipment, involving excessive waiting times and all too often associated with a high mortality rate through poor hygiene and late referral – deter people from seeking MCH care. The interlinked issues of acceptability, accessibility and underutilization provide the necessary criteria to judge the effectiveness of the individual elements of GOBI–FFF and the overall performance of MCH strategies.

Growth monitoring

As over a quarter of the children in the developing world suffer from invisible malnutrition, the use of the 'Road to Health' chart (see Morley and Lovel 1986) to make this problem visible to mothers could potentially have an extremely powerful impact. The chart is an invaluable, simple and low-cost means to measure and monitor the normal rate of growth of a child. It is not simply a recording device but a practical, powerful educational tool with a main message that monthly weight gain is good, constant weight is a warning, and weight loss is a danger sign. As research in Ghana has shown, growth monitoring technology needs to be introduced into communities in appropriate and acceptable ways, with due recognition of local beliefs and behaviour patterns regarding child growth. For an improved understanding of growth charts and an appreciation of their value and significance by the whole community can lead to an increase in their generalized use and subsequent effectiveness.

Oral rehydration therapy

Oral rehydration therapy (ORT), which can be said to be potentially the most important medical breakthrough this century, is simple and affordable and has been proven a safe and effective method of preventing and correcting dehydration caused by diarrhoeal infections. Such dehydration is estimated to kill approximately 3.5 million children every year, with bottle-fed infants and children undergoing weaning being particularly susceptible. Once restricted to medical intravenous use only, and physically and financially beyond the reach of most people, ORT is now available in solution form

Table 9.1 The ORS debate

	Home-made solutions	Prepared packaged solutions

Advantages:	Cheap
	Practical/easy to understand
	Uses local, readily
	   available resources
	   (local sugar substitutes, e.g.
	   coconut water,
	   appropriate measures, e.g.
	   home-made spoons,
	   finger measurements)
	Encourages self-sufficiency
	   and self-care
	Control in hands of family
	Emphasis on the preventive
	Early treatment at first
	   signs of diarrhoea
		Immediate impact
		'Magical' results
		Rapid acceptance
		Precise, exact measurements
		   – known mixture
		Emphasis on the curative

Disadvantages:	Imprecise measurements
	Greater room for error
	Time-consuming application
		Costly – reaches few people
		Supplies often uncertain
		   and inadequate
		Stop-gap measure
		Insufficient production to
		   reach all children with
		   diarrhoea
		Greater dependency on health
		   professionals and high
		   techology, and on external
		   resources subject to both
		   national and international
		   politics
		Poor local understanding

that can readily be administered by mother or child-carer. However, limited availability and usage have diminished its immediate global impact. Levels of acceptance need to be greatly improved upon if the year 2000 target, use by 50 per cent of parents in developing countries, is to be met.

While the formats are not mutually exclusive, considerable debate has centred on the more appropriate format of oral rehydration salts (ORS) to promote, whether in commercially prepared sachets or in home-made solution. The issues, as summarized in Table 9.1, revolve around the financial costs, the degree of control (whether self-determination or external reliance), the concomitant levels of understanding, an emphasis on prevention or cure, and the durability of the strategy (whether a long-term measure or merely a stop-gap).

ORT is theoretically sound, but does it work in practice? Its effectiveness depends upon increasing awareness and widespread acceptance at the local level. The successful promotion of ORT in Honduras through an intensive multi-media communication campaign illustrates the potential impact of such a strategy, as infant deaths from diarrhoea fell by 40 per cent within a two-year period. However, by describing ORT as a mass 'solution', it has been strongly argued that too much emphasis has been placed on the dangers of dehydration and that the root causes of diarrhoea – malnutrition, poor sanitation and various infections – have been forgotten. Contaminated water supplies need to be rectified, malaria identified, reliance on bottle-feeding reduced, and levels of general environmental hygiene improved. But as ORT is currently a separately funded 'vertical programme', integration into a comprehensive PHC approach will prove difficult.

Breast-feeding

Breast-feeding is perhaps the most effective measure of all the GOBI–FFF interventions, as its key role in each of the other MCH priorities underlines. It is not only important for the growth and nutrition of the infant, but can prevent diarrhoea. Breast milk can provide a form of passive immunization against infection, while breast-feeding is also a natural form of family planning. It is also rather obviously linked to feeding practices. Although these benefits have long been recognized, over the last decade breast-feeding has experienced considerable decline. It is becoming taboo by modern convention. Growing urbanization, with its concomitant absorption and adherence to 'modern' tastes, has encouraged mothers in the uptake of status-giving bottle-feeding. Bottle-feeding threatens the lives of millions of children. It is expensive, and bottle-fed babies are more likely to be malnourished and to contract diarrhoeal infections, through milk powder being overdiluted with unclean water in an unsterilized feeding bottle that is often left to stand in the tropical heat. All too often, the safest and most nutritional source of milk is neglected. The breast is best.

Campaigns to defend and promote breast-feeding, and to stop the spread of artificial substitutes, gathered momentum during the 1980s. International effort targeted the damaging marketing practices of infant formula manufacturers and supported worldwide boycotts of company products, while programmes advocating the advantages of breast-feeding have sought to increase awareness among women. However, if such programmes are to succeed, mothers need the necessary legislative and technological support. Issues of job security and paid maternity leave, the provision of nursing breaks and child-care facilities, and technologies to lighten women's workloads all need to be addressed. But most important of all, increased awareness and support from men could really improve maternal and child health. The understanding that pregnant and lactating women need to work less and to eat more, and men's greater involvement in the raising of their young, would make a significant improvement to the lives of women and children.

Table 9.2 The immunization scenario

	Benefits	Obstacles
Medical and technological:	Clear-cut intervention, immediate impact Vaccines are safe, simple, easy to use, comparatively cheap, effective (reduce mortality and morbidity) and reduce caseload on limited curative health resources Costs less than prolonged treatment Protection Vaccines retain potency for longer periods and at higher temperatures 'Universal spare parts kit' Solar-powered refrigerators Colour-change temperature warnings Low-cost sterilization techniques	Difficulties in supplies, storage and logistics influence vaccination coverage and quality of immunization Insufficient, limited financial resources Funding for capital budgets (fridges, vehicles) is readily available, but long-term recurrent expenditure (spare parts, wages, fuel, vaccines) is limited/lacking Breaks in cold chain operation/ equipment Reaching selected high-risk population groups – poorest of poor, nomads/pastoralists, seasonal migrants
Social:	Benefits readily apparent Little change of behaviour needed Immediate savings apparent Good value for money Highly 'visible' results High-profile political support Improved awareness: health education campaigns, social marketing, leading to greater community involvement, resulting in increased demand and uptake	Lack of knowledge results in low use of immunization services Organizational problems Short-term mass campaigns – lack of means for sustained effort, no continuity of contact, service taken for granted once initial enthusiasm ends

Immunization

Measles, whooping cough, tuberculosis, diphtheria, poliomyelitis and tetanus – the 'big six' killer diseases – have been estimated to account for approximately 5 million child deaths a year and to disable 5 million more children. Thankfully, such a tragedy is now preventable using newly improved vaccines. In 1978, following the success of national immunization campaigns (with coverage levels of 50 per cent) and the eradication of smallpox, UNICEF spearheaded the establishment of the Expanded Programme of

Immunization (EPI) with the goal of universal child immunization by 1990. Perhaps such a goal was too optimistic. Coverage of 90 per cent has been achieved, but in general rates still vary drastically both between and within countries.

Immunization campaigns can seem an attractive prospect and their undoubted benefits are readily apparent. However, if universal coverage is to be achieved, then a complexity of problems and obstacles, which often go unappreciated, need to be overcome. Social breakthroughs are just as important as technological progress (Table 9.2). The availability of vaccines and their efficacy, safety and ease of use have been the focus of much attention, along with major investment in the development of 'cold chains' – the unbroken relay of transport and storage, which must maintain vaccines at the right temperatures from the point of manufacture to the point of use.

Levels of acceptability, accessibility and utilization are crucial to the success of immunization campaigns. Merely making immunization services more readily available does not necessarily increase acceptability. The behavioural aspects of immunization and high drop-out rates have become matters of serious concern. Immunization cannot be viewed as purely as 'technological fix'; people's attitudes and beliefs are of great consequence. Low acceptance and uptake rates may be due to a poor immunization system with irregular, inconvenient clinics, staff and equipment shortages, high costs and long waiting times. This may be further exacerbated by poor dissemination of information regarding the whole immunization programme, which serves only to reinforce unfavourable attitudes towards immunization. Faith in EPI services diminishes rapidly with concern about the unexplained adverse effects of vaccines, and with knowledge of immunized children who have failed to be protected adequately. Increased accessibility is equally important. To reach the 'high-risk' groups – the poorest of the poor – socio-economic barriers and problems of poor transport and communications, together with illiteracy, need to be addressed. Mobile outreach services have been particularly successful in reaching scattered, isolated populations, but these are expensive. It has been argued that a combination of static and mobile services, mass campaigns and the increased involvement of unofficial health service resources, such as traditional birth attendants (TBAs), could be effective in increasing acceptability and thus improve immunization coverage further.

The two central concerns of the EPI, namely its pure selectivity (the concentration on specific ailments) and the ethical question of the imposition of technology, have been the focus of much criticism. Immunization is a short-term measure that fails to address the underlying problems. It is a rather sobering point, but, despite the immediate reduction in infant mortality following immunization, the net gain in child survival is very small. Immunization cannot be seen as a magical solution. It merely delays mortality. Children die instead from malaria, diarrhoeal diseases and acute

147

respiratory infections (ARI). This tendency to concentrate upon a number of specific causes of child death has diverted attention and drained resources from other equally important aspects of MCH care. Critics further argue that the imposition of such 'high-tech' interventions only exacerbates the existing dependency syndrome, encouraging further reliance upon Western countries' funds, vaccines and equipment, and requires little active participation from people. Communities are not consulted as to their wishes. Immunization has become an end in itself. In their defence, advocates of the EPI strongly argue that immunization has proven cost-effective; the cost per person, a nominal $1.50, clearly underlines the utility of the programme. The cost of the first day of the Gulf War alone could have immunized 60 million children. UNICEF counters all criticism, and ends all debate, with the argument that people today find it unacceptable to allow children to die from what are *preventable* causes – 'a stain on our civilization'.

THE THREE Fs

By the mid-1980s growing recognition that the health of mother and child were inextricably linked, and thus the realization that any strategies targeted solely at children could have only limited impact, led to 'three Fs' being added to the GOBI strategy: Food supplements, Family spacing and Female education. Although more deep-seated and therefore more difficult and more costly, such was the potential significance of these changes that it was estimated that child mortality and morbidity could be reduced by two-thirds within a decade.

Food supplements

Low birth-weight is one of the most important causes of infant deaths in developing countries. It accounts for between 12 and 15 per cent of births, and between 30 and 40 per cent of infant deaths. Underweight babies are three times more likely than other babies to die in infancy, and the survivors are more susceptible to longer and more frequent illness. The nutritional status of the mother is crucial. 'Full-term' low birth-weight is a direct result of too little food and inadequate nutrition in pregnancy, compounded by excessive hard physical work, and exacerbated by anaemia and successive births. Selective supplementation during pregnancy of iron, folic acid, protein and calories could reduce the number of cases of low birth-weight by around 60 per cent. However, to date, such programmes have not been particularly successful. They have been criticized for being expensive, for requiring sophisticated logistics to transport food supplies to remote places, and for fostering a passive 'hand-out' mentality rather than helping to develop self-reliance. Many argue that food supplementation should be solely restricted to emergency situations. To ensure affordability and cost-effectiveness, food supplementation programmes need to be targeted at those individuals most in need, and to be based upon locally available food supplies.

Family spacing (family planning)

Birth spacing is one of the most powerful ways of improving the health of women and children. Births which are 'too many or too close', or to women who are 'too old or too young', are responsible for about one-third of all infant deaths worldwide. The World Fertility Survey (WFS) concluded that, if all births were spaced at least twenty-four months apart, then up to 50 per cent of infant deaths in developing countries could be avoided. Child spacing, regulating the number and timing of births through family planning, has the potential not only to reduce infant mortality and morbidity, and to significantly influence the mental and physical development of children, but also to enable women to control their own fertility, and thus reduce the potential health risks of pregnancy and childbirth.

Evidence from the WFS has revealed that demand is there for family planning services, but problems of acceptability and uptake hinder their success. To be fully effective, family planning (FP) needs to be sensitive to the social and cultural climate, and to take into account local customs and beliefs surrounding sexual matters and attitudes towards fertility. However, it is still a particularly sensitive issue, subject to suspicion and misunderstanding. Accusations of fertility control for political gain have led to resentment and rejection. But most resistance to FP is largely through ignorance; therefore an increased knowledge and improved awareness of FP programmes is vital to allay such fears.

Social, political and religious constraints are further heightened by problems of contraceptive availability and the accessibility of FP services. In the 1970s community-based delivery systems were developed to demystify FP and put it at the disposal of millions. Local people were given brief training to provide basic FP services in their own communities. To improve access by overcoming geographical, cost, cultural and communication barriers, 'social marketing' has been applied to contraceptive use in a similar fashion to ORT and immunization campaigns. Contraceptives have been made readily available through retail outlets – pharmacies, corner shops, cigarette kiosks and vending machines – at subsidized prices and promoted like any other commercial product through advertisements in the mass media. In some countries FP has been given a high profile and made a political priority; consequently its practice has spread rapidly. Such social marketing of FP over the last two decades has given rise to great hope for the whole 'child survival and development revolution'.

Future strategies need to ensure that FP is made yet further available, so that all individuals have the wherewithal to control their own fertility according to their own wishes and interests. FP services need to be more closely integrated with MCH care, as several contraceptive methods require medical intervention, and pre- and post-natal clinics already bring women into contact with health services, thereby providing good counselling opportunities. Since FP and MCH are mutually reinforcing, in the long term family planning cannot be divorced from health development in general.

Female education

The importance of maternal education to child survival is now widely recognized. Both infant mortality and birth rate decrease with improvements in female literacy. Educated mothers may attach higher value to the health and welfare of children, may be less fatalistic about disease and death, may be more knowledgeable about disease prevention and cure, and are more likely to adopt new codes of behaviour. Thus education works as 'medication against fatalism'. Unfortunately in many parts of the world, female illiteracy rates remain as high as 85 per cent, and considerable resistance persists towards female education. Poor educational levels reflect and perpetuate women's low status. Although change is slow, the proportion of literate mothers is increasing through the promotion of non-formal literacy programmes. For MCH care this is particularly significant, as education increases women's access to vital information and their confidence to put new ideas into practice. An increased awareness of the benefits of growth monitoring, ORT, breast-feeding, immunization and family planning can only enhance the effectiveness of these strategies.

THE SAFE MOTHERHOOD INITIATIVE

The significance of maternal health has until very recently been accorded a low priority. Traditionally, the improvement of women's health has been viewed largely as a vehicle to improve child health, rather than as an important objective in its own right, for its own intrinsic value. Since the United Nations Decade for Women (1976–85) the full magnitude and sheer scale of the suffering associated with maternity have become apparent. For a woman in the developing world, the average lifetime risk of dying from a pregnancy-related cause is around 1 in 15 – so commonplace that it is accepted as a 'normal' part of life. More recently, an increased recognition that women need to be fully integrated into development efforts has led to the realization that this 'neglected tragedy' could be prevented. Low-cost and effective interventions were available. No technological breakthroughs were needed, just the will to give it priority.

The advent of the WHO's Safe Motherhood Initiative (SMI) in 1987 did just that, setting itself apart from other health initiatives as the first to focus specifically on the health of women. The SMI advocated a combination of health and socio-economic strategies to improve the quality and longevity of women's lives, urging action to address the immediate and underlying causes of maternal mortality and morbidity. The target was to reduce maternal mortality by at least half by the end of the century. The World Bank estimated that the mere investment of US$2 per person per annum could achieve this.

Priority was given to improving maternal nutrition and women's general health status rather than concentrating exclusively on the mechanical aspects of obstetrics. The initiative endorsed the PHC community-based approach. In a four-pronged strategy, special emphasis was placed on: raising women's socio-economic, legal and educational status, based on a more equitable distribution of resources and workload; increasing women's access to and use of family planning services, in order to change reproductive behaviour, minimize the risk of unwanted pregnancies and thereby reduce the number of illicit and unsafe abortions; improving routine care during pregnancy and delivery, through pre-natal supervision and treatment stressing early and regular attendance, and preventive strategies – screening to identify mothers at risk, the use of iron and folic acid supplements and anti-malarials, and the establishment of 'maternity waiting homes', plus simple but experienced care at delivery by trained traditional birth attendants (TBAs); and lastly improving the quality and accessibility of treatment for obstetric emergencies, through rapid, efficient referral and an upgrading of services at the district level to cope with minor maternal surgical procedures and blood transfusions.

Much discussion within safe motherhood programmes revolves around the relative importance of the various interventions and, as ever, where to target limited resources. But if the SMI is to succeed fully it needs to be implemented as a balanced, linked system so as to maximize the combined potential of these strategies.

THE EFFECTIVENESS OF MCH STRATEGIES

Questions have increasingly been asked of the value of MCH strategies, as high levels of maternal and child mortality and morbidity still characterize the health profiles of most developing countries. If future priorities are to be set, then the effectiveness of these short-term solutions needs to be considered, for such vertical interventions run contrary to the comprehensive PHC approach. UNICEF has argued that, since resources are never adequate, priorities have to be established, and thus a strategy of selective primary health care (SPHC) would serve as a starting-point that could consolidate and grow into comprehensive PHC.

The selective–comprehensive debate has yet to be resolved. Advocates argue that GOBI occupies a special place in PHC. The benefits are fourfold: the strategies use appropriate technologies, widely available at low cost; they produce tangible results which can lead to acceptance of the wider cause of PHC; they empower parents actively to protect their children's health, promoting confidence and self-reliance; and they encourage participation rather than passive recipient dependence, thereby promoting the very essence of the PHC idea. Critics of SPHC contest this, arguing that it is more effective in the long run to tackle, from the very outset, the root causes – population control through family planning and improvements in poor environmental conditions, inadequate nutritional standards and women's low social and economic status – rather than to target the specific direct causes. The issue of dependency on technocentric solutions that encourages reliance on overseas imports of drugs and vaccines, and sustains the vested interests of the developed world, remains a strong philosophical objection.

If these selective strategies are to become acceptable and effective in coverage in the future, then much can be learned from the past experience of industrialized countries earlier this century. Improvements in socio-economic status and living conditions resulted in steady reductions in infant and child deaths; but maternal mortality remained largely unchanged until it declined sharply following the introduction of widely available antibiotics in the 1950s and the development of specific high-tech interventions to treat obstetric complications. Such divergent paths underline the need for a broad-based PHC framework for grass-roots development. With preventive health measures practised alongside curative medicine, and operating in parallel to programmes in water and sanitation, in environmental health and in housing, PHC will thus remain comprehensive in aim and targeting those at risk.

DIRECTIONS FOR THE 1990s

The early 1980s saw a determined effort through specific MCH strategies to improve the situation of mothers and children. Their powerful impact gave hope to the promise that the 'child survival and development revolution' was possible, and that the United Nations' goal of reducing infant deaths to below 50 per 1,000 births in all nations by the year 2000 could be achieved. However, a continued world economic recession, the impact of the growing debt crisis and the advent of structural adjustment packages have threatened to undo the progress and advances of the 1980s.

It is now well recognized that the effects of this harsh decade weighed most heavily on the poor and those most vulnerable – namely mothers and children – who experienced greater poverty and a deterioration in their health status. The incidence of low birth-weight, a sensitive indicator of the well-being of women, increased; and, following the removal of food subsidies, UNICEF studies revealed a widespread decline in the nutritional levels of children and pregnant and lactating women in both rural and urban settings. The adverse economic climate led to substantial cuts in public expenditure on health services, resulting in shortages of drugs and essential medical supplies, and of transport and fuel. The effectiveness of various MCH strategies significantly deteriorated. Coverage rates of national EPI programmes dropped, and diseases virtually eradicated by past campaigns reappeared as major epidemics. In many poor rural communities PHC was no longer adequately functioning. Differential user charges were introduced, with a consequent decline in clinic attendances. One practical response, the Bamako Initiative, sought to buy essential drugs in bulk at low cost, and sell them at prices affordable but sufficient not only to replenish stocks but also to provide the running costs of local MCH services. Such was the impact of economic forces on mothers and children that UNICEF called for 'adjustment with a human face', with social expenditure on education, health and sanitation to be given a higher priority and arguing that protection for the most vulnerable should be foremost.

However, issues of debt and adjustment to economic recession are likely to remain major influences on the lives of mothers and children throughout the 1990s. The know-how and low-cost technologies are available for preventing and treating at least three-quarters of all maternal and child deaths. The need now is therefore to consolidate and build upon the progress of the 1980s, to achieve the targets for the year 2000 as set out by the 1990 World Summit for Children.

The Summit succeeded not only in putting the survival, protection and development of children firmly on the political agenda, it advocated long-term investment in people and their health, nutrition and education to foster economic progress, and promised a new political commitment. Such notions were translated into specific goals for the year 2000, including a one-third reduction in child deaths, 90 per cent immunization coverage, a halving of child malnutrition and maternal mortality, universally available family planning, and basic education for all children. The Summit further reiterated the critical role that women play in the well-being of children, and drew special attention to their health, nutritional and educational requirements from childhood, with access to pre-natal care, to trained assistance during childbirth and to referral facilities for high-risk pregnancies and obstetric complications given strong emphasis. Thus maternal mortality had at last been made a high priority in the strategy for 'Health for all by the year 2000'.

The launch of a number of new initiatives and technological break-throughs during the 1990s may also greatly help to realize these goals within the decade. In 1991 UNICEF and the WHO launched the 'baby-friendly hospitals' campaign to promote the practice of breast-feeding, by focusing on nursing practices to significantly influence maternal behaviour. Provision is currently under way for hepatitis B to be included as the 'seventh vaccine' in all national immunization programmes, and research on the children's vaccine initiative (CVI) is being stepped up. This initiative aims ultimately to extend the range of affordable vaccines to include ones against malaria, respiratory infections, meningitis, influenza B, dengue fever and AIDS; to develop more heat-stable vaccines that remove the necessity for refrigeration and thus ensure widespread use; and to incorporate several vaccines into one carrier and for their contents to be released gradually over time. The ultimate goal is to be able to administer all of today's vaccines in a single injection – a single shot 'super-vaccine'.

One further objective for the 1990s, and the focus of much discussion, is the need to develop integrated MCH services in which curative, preventive and educational services would all be available on a single visit, thereby avoiding multiple visits for ante-natal and under-5s clinics, and overcoming many of the accessibility barriers that women face. The instigation of such a 'family care' system could have enormous impact on maternal and child health.

THE IMPORTANCE OF MCH STRATEGIES TO DEVELOPMENT

To ensure continued support for MCH strategies in the 1990s and to achieve substantial social progress by the end of the century, UNICEF called for renewed impetus in the promotion of 'real development'. This new focus for development efforts concentrates on the poorest of the poor, aiming to enhance their abilities to control their own lives. 'Competent health care' prioritizes pregnant women and children simply because they are the most vulnerable, and demands the protection of children's physical and mental development, giving them 'first call' on societies' resources.

By investing in people and by satisfying their basic health and education needs, a 'market-friendly' or human-centred approach to development has been advocated as one of the most effective means of stimulating long-term economic growth and thus improving general welfare. MCH strategies provide the experience and low-cost opportunities for relatively small investments to lead to considerable social and economic returns. At a time of fierce competition for resources, such strategies must be realistically achievable on the large scale, be of low cost and high impact, and be politically attractive. If they are to accelerate real development and double the cost-effectiveness of development efforts, then their sustained coverage and expansion must depend upon participation and mobilization – putting essential health knowledge at the disposal of the majority.

However, the full potential of these strategies will never be fulfilled unless the issue of 'apartheid of gender' is addressed. Women play a pivotal role in development, yet their needs are often ignored and their potential under-utilized. Their disproportionate contribution is all too often rewarded only by the 'disease of discrimination'. This undervaluing of women is seen as a major obstacle on the road to an equitable and sustainable future.

Improvements in their health and overall status are preconditions to ensure their active participation in the development process.

Efforts to reduce maternal and child deaths and ill-health are crucial not only for their own sake but as an important means to slow population growth and thus ensure environmentally sustainable development – the two central issues on the agenda for the 1990s. The link between fewer deaths and fewer births is one of the least understood but most vital of contemporary concerns. Lowering child deaths helps to lower birth rates, as fertility rates tend to drop when parents are more confident that their children will survive. Well-informed planning of births is one of the most effective and least expensive ways of improving the general quality of life. Thus family planning, improved status and self-determination can give women significantly more control over their lives, which is after all what real development is all about.

But perhaps the greatest contribution that MCH strategies can make to the overall development process lies in their ability to act synergistically. Each of the various strategies can have a considerable impact. However, this impact is greatly magnified once each of the basic elements enhances and multiplies the effectiveness of the others. For example, female literacy aids family planning programmes; fewer pregnancies improve maternal and child health; better health improves educational performance; improved schooling leads to increased agricultural productivity and higher incomes; improved incomes in turn benefit diets, child health and survival rates; fewer child deaths help lower birth rates; and smaller families consequently mean healthier mothers and children. Thus these components of real development are linked together in a mutually retarding or mutually reinforcing relationship, in which the whole is very much greater than the sum of its parts. Such a capability underlines the importance of MCH strategies to realize social progress and an improved quality of life throughout the developing world, and clearly illustrates their combined potential to achieve the WHO's 'Health for all by the year 2000'.

SELECTED READING

Ebrahim, G.J. (1982) *Child Health in a Changing Environment*, London: Macmillan.
Morley, D. and Lovel, H. (1986) *My Name is Today*, London: Macmillan.
Royston, E. and Armstrong, S. (eds) (1989) *Preventing Maternal Deaths*, Geneva: World Health Organization.
Smyke, P. (1991) *Women and Health*, London: Zed Books.
United Nations Children's Fund (annual) *The State of the World's Children*, Oxford: Oxford University Press.
Williams, C.D., Baumslag, N. and Jelliffe, D.B. (1989) *Mother and Child Health: Delivering the Services*, Oxford: Oxford Univesity Press.

10

SOCIO-ECONOMIC CHANGE AND THE HEALTH OF ELDERLY PEOPLE

Future prospects for the developing world

Anthony M. Warnes

INTRODUCTION

This chapter examines the interplay between socio-economic change or 'development' in the low-income nations of the world, their current demographic trends and the material, welfare and health situation and prospects for their older people. It aims to elucidate the complexities of the interactions and go beyond a superficial presentation of the consequences of demographic ageing. To do this, basic facts and figures are presented, with guidance as to where more detailed information may be found. Then the health and welfare situation of elderly people is considered, particularly in the light of current trends in urbanization and economic change. The final sections set out some of the policy options that face governments and public bodies in the developing world.

There are many definitional, conceptual and methodological problems in discussing these issues. We will have to proceed with caution, but on the other hand we cannot delay too long over these refinements. A relatively broad view of health, incorporating morale and economic well-being, will be adopted. Development is seen as the manifold social and societal changes that have been widespread in recent decades, and not just economic growth. And elderly people are taken as those who have attained a minimum age, generally 60 years, although some international data sets and commentaries which are referred to use 55 or 65 years. But the most important caveat is that our knowledge of the health of elderly people throughout the low-income nations is at best impressionistic. The few excellent local surveys highlight both the problems of defining and measuring health and our ignorance of more general conditions (Andrews *et al.* 1986).

Table 10.1 The elderly population (65+ years) of less developed world regions, 1960–2020

	Population (millions)			Share of total population (%)		
	1960	*1990*	*2020*	*1960*	*1990*	*2020*
More developed nations	80	146	232	8.5	12.1	17.3
Less developed nations	79	180	471	3.8	4.4	7.0
Africa	8	19	53	3.0	3.0	3.7
Caribbean and Central America	2.5	6.5	17	3.6	4.6	6.9
South America	5	15	37	3.5	5.0	7.8
South-east Asia	7	17	46	3.3	3.9	6.9
South Asia	21	46	129	3.5	4.1	6.3

Source: Keyfitz and Flieger (1990)

THE MECHANISMS OF DEMOGRAPHIC AGEING

The population of the less developed nations has been increasing very rapidly in recent decades. More children were born during the 1960s than in the 1950s, and more still during the 1970s. As future decades pass, so therefore will greater numbers of people reach 60 years, and the absolute size of the elderly population will grow (Table 10.1). However, during the last two decades, falling birth rates have been spreading among the less developed countries, particularly in Latin America and South-east Asia. The increase in the number of births is slowing down, and on best current estimates the annual number of births will peak around the end of this century at about 126 million (Keyfitz and Flieger 1990: 107). Well before then, the population pyramids of less developed nations will have reflected the fall in fertility, with their bases becoming progressively less broad (relative to the higher and older age groups) and their sides less steep.

A sustained decrease in the birth rate brings about *demographic ageing*, the progressive increase in the average age of the population and in the share that is elderly, defined as the people who have exceeded a certain age, whether 50, 60, 65 or 75 years (Myers 1990). Actual population data show the beginnings of the age-structure change in many Third World countries; and where detailed population projections are possible, a rising momentum is predicted. Among the world's continents, only in Africa are few countries yet giving evidence of the change. Using the United Nations demographers' division of the world into the more developed and less developed regions, in the latter the population aged 65 or more years has increased during the last thirty years by 101 million or 128 per cent (from 78.8 to 179.8 million during 1960–90) (United Nations 1985). Their share of the total population has increased from 3.8 per cent to 4.4 per cent. During the next thirty years the 65-plus years population is likely to grow by nearly 300 million or 162 per cent (to 470.6 million in 2020), and the population share to grow further

157

to 7 per cent. A further demonstration of the growing momentum of demographic ageing in the developing world is that its proportion of the world's elderly people has increased from one-half in 1960 to 55 per cent in 1990, and it is likely to be two-thirds by 2020 (Keyfitz and Flieger 1990). The average age of the population of the less developed world was falling during the third quarter of this century, from 25.3 years in 1950 to 24.7 years in 1980, but it is now increasing and is likely to do so with accelerating pace to reach around 31.1 years by 2020.

Demographic ageing will be fastest where fertility is falling most rapidly. There is of course great variation among the nations of the world in the pace of demographic ageing that they are experiencing. This cannot be described in full detail, because population projections require particularly reliable and detailed data on age-specific birth and death rates, and these are not available for all countries. One recent review, however, suggested that during 1988–2020 there would be more than 200 per cent increases in the population aged 55 years or more in four Asian countries (Indonesia, Singapore, Philippines, Thailand), two African countries (Kenya and Zimbabwe) and four Latin American countries (Costa Rica, Mexico, Brazil and Peru) (Kinsella 1988: Fig. 2).

The imminence of rapid demographic ageing has captured the attention of many international agencies and national governments. Its recognition brings deserved scrutiny but has also resulted in two simplistic and contradictory first interpretations. On the one hand, alarm is raised about the prospective growth of 'the elderly', who are assumed to be economically unproductive and also great consumers of health and welfare services. On the other, it is reasoned that because elderly people are increasing absolutely and relatively, individuals must be living longer, and therefore at any given age, say 60 or 70 years, the prevalence of ill-health must be falling. Both inferences are largely reflections of interpretations for the more advanced stages of demographic ageing in the industrialized nations. They do not stand up to critical examination for the less developed realm.

THE URBAN–RURAL DIMENSION

In several less developed countries, and characteristically in Latin America, the decades of highest birth rates and most rapid population growth also saw very high rates of urbanization. When used in its strictest sense, this term indicates a relative transfer of the population from rural to urban habitats, i.e. it implies more than urban population growth and usually a combination of net migration from rural to urban areas and a higher urban natural increase rate (recently most often from lower infant mortality). Third World rural–urban migration exemplifies the universal tendency for adolescents to dominate migration flows. The result has been an accumulation of young fertile adults in rapidly growing cities and a tendency for rural

Table 10.2 Projections of the elderly population (60+ years) of the Bangkok Metropolitan Area, 1990–2020

	1990	2000	2010	2020
(A) *Constant mortality schedules and migration schedules constant*:				
Population 60-plus years (00,000s)	3.6	5.6	8.4	15.1
Share of the total population (percentage):				
Migration decreasing	5.7	7.0	8.8	13.3
Migration constant	5.7	6.9	8.6	12.8
Migration increasing	5.7	6.8	8.4	12.1
(B) *Mortality decreasing by 1% per annum and migration schedules constant*:				
Population 60-plus years (00,000s)	3.6	5.8	9.1	16.9
Share of the total population (percentage):				
Migration decreasing	6.9	7.4	9.9	16.2
Migration constant	6.9	7.3	9.7	15.6
Migration increasing	6.9	7.3	9.5	14.8

Source: Warnes (1992)

Notes: Migration schedules are the age- and sex-specific rates of net migration between Bangkok Metropolitan Area and the rest of Thailand. Decreasing and increasing are at 1.5% per annum.

populations to have higher elderly shares than the cities (Kinsella 1988). But this differential is almost certainly ephemeral and about to be reversed.

The rapid-growth cities now tend to have overrepresentations of people aged 20–50 years, and commonly their age structures already show a greater recent decrease in fertility than rural areas. In many parts of Latin America, North Africa and South-east Asia, the smaller families and households of the urban populations, and their higher educational attainment and female (paid) labour force participation rates and greater life expectancies, distinguish them from their rural cousins as being more 'modern'. The history of urban population growth means that the large city populations of less developed countries will imminently experience exceptionally rapid change in their age structures. Within twenty years, the high birth cohorts of the 1960s will reach old age. Few of them will return to rural areas, and probably only a minority will have contributed to traditional family occupations.

Detailed projections have been computed for the Bangkok Metropolitan Area (BMA) of Thailand, with various assumptions as to the future trends in fertility, mortality and net migration transfers between the city and the rest of the country (Warnes 1992) (Table 10.2). The base data for the projections are the detailed birth, death and migration schedules from the 1980 national census. All variant projections suggest that in 1990 around 5.7 per cent of the BMA population was aged 60 years or more. All variants also suggest a very rapid increase of this share, particularly after 2010. If mortality and migration rates are constant, the elderly population of the BMA is likely to increase from 360,000 in 1990 to 1.51 million in 2020, that is, to around 13 per cent of the metropolitan total. If mortality rates (at all ages) fall by 1 per cent each year, then the 2020 elderly population will be even higher, at 1.69 million or 15.6 per cent of the total. And finally, if migration rates begin to moderate, as might occur if the nation adopts successful policies to decentralize new economic developments, then the rejuvenating effect of young in-migrants will be lost and the pace of ageing will be even faster.

Comparable ageing projections seem likely for many other rapid-growth cities of the Third World. The 'bright-lights' effect of the 1960s and 1970s which attracted so many young people will produce an 'evening shadow' in the early twenty-first century, and thereafter it will lengthen as long as birth rates decline. The large increments of elderly people will include progressively fewer who can turn to traditional modes of support in old age. Some decades of unprecedented disruption to the lives of elderly people, of stress and deficient support, are possible. On the other hand, at least in the economically more successful nations and among the more educated groups, there will be increasing numbers with more than minimal savings, a dwelling, assets besides and pensions entitlements. In this way, the adjustments to demographic ageing faced by less developed nations will probably be highlighted in the cities. They will have faster and more visible changes, in both their age structures and their forms of economic activity and income generation. We should therefore be critical of the impression which is gained from national projections over the next thirty years. For most countries they suggest a slowly gathering pace of ageing, but for the major metropolitan areas a much more rapid progression to a near-European elderly population share is likely.

THE 'EPIDEMIOLOGICAL TRANSITION' AND MORTALITY AND HEALTH IN OLD AGE

Demographic ageing could theoretically occur without any change in the life expectancy of an individual at 1 year of age – it does not necessarily imply that people are living longer adult lives. In actuality, in many less developed countries survival and health status have been improving in middle and old age as well as around childbirth but at a much slower rate. One nation with excellent life table information is Malaysia. Between 1970 and 1985, infant and child mortality rates fell massively, with a 69 per cent improvement for girls aged 1–4 years (Keyfitz and Flieger 1990). At older ages, improvements have been less. For those aged 60–65 years, a one-fifth decline was achieved. For those in their early seventies, the improvement for females was 9 per cent but for males only 1.2 per cent; and for those in their early eighties, death rates increased. Similarly in Puerto Rico during

1970–85, while the infant mortality rate declined by one-half for boys and by one-third for girls, increasing age was associated with progressively lower mortality improvements. For those aged 80–84 years, among men the improvement was only 1.8 per cent, and among women mortality rates increased.

In Thailand also, the substantial falls in infant and child mortality have not been matched by the rates for those aged over 55 years. Cause-of-death data help little in understanding this rigidity, partly because old age and other ill-defined conditions constitute most of an 'others' category which accounts for two-thirds of all deaths. Through the early 1970s, heart and cancer conditions were diagnosed for less than 6 per cent of all deaths. There is, however, much avoidable ill-health and mortality from infectious diseases, with leading causes being respiratory and digestive-tract conditions. These could be controlled by changing habits, immunization and improved sanitation (Warnes 1992). A recent careful examination of the life tables of Thailand concludes:

> it is difficult to come to any firm conclusion about recent trends in older age mortality in Thailand. While there is some reason to believe it has improved over the last two decades, especially for women, uncertainties about data quality and . . . computational methods leave doubts about what actually has been the case.
>
> (Chayovan *et al.* 1990: 10)

Earlier chapters have outlined the proposition that development is accompanied by an epidemiological transition. This represents a shift in the predominant causes of death away from infectious, parasitic and nutritional diseases and towards the degenerative disorders associated in the rich world with old age. As the process of demographic ageing has only just begun, so also most less developed nations are only at the first stage of the epidemiological transition. Perhaps one should add, 'as far as we know', for despite the best efforts of world organizations and of scientific medicine, identifying the causes of deaths is by no means straightforward. Most non-accidental or non-violent deaths represent the inability of the body's protective mechanisms to combat a range of attacks and insults. Many deaths arise from multiple causes, and this is particularly likely where malnutrition is widespread, and minor digestive infections commonplace. A detailed autopsy is rarely conducted; in all countries cancers or neoplasms are particularly difficult to identify without investigation; and social and cultural acceptability influences the naming of causes of death.

These difficulties do not disguise the probability that much of the changed balance of causes of death to date in less developed nations can be accounted for by the massive and magnificent reduction of infant mortality. In Mexico, for example, infant mortality rates have fallen from around 325 per 1,000 at the beginning of this century to about 216 in 1925, 96 in 1950, 53 in 1975 and 42 in 1983. It is easy to understand the changing causes of death in the first stage of the epidemiological transition if the main effect is such an enormous reduction. If we assume that infant deaths drop from around 35 per cent to 5 per cent of the total deaths and that none of them is the result of degenerative causes, and then further assume that among the deaths after 1 year of age 70 per cent are attributable to the degenerative causes, then the assumed fall in infant mortality would increase the proportion of deaths from degenerative causes from 46 per cent (70 per cent of 0.65) to 66 per cent (70 per cent of 0.95). This occurs with no change in the morbidity and mortality of old people. Phillips (1990, 1992) reports that by 1988 in Hong Kong 58 per cent of all deaths were ascribed to neoplasms or cardiac and cerebro-vascular disorders.

Population surveys and hospital and primary care records are the principal sources of information about the health of elderly people. Inevitably their coverage is limited, and repeat surveys or longitudinal comparisons are rare. Thailand has unusually full data (Kinsella 1988: 14–15; Chayovan et al. 1990; Warnes 1992). Successive government health and welfare surveys since 1974 suggest a halving of the prevalence of illness (in the previous two weeks) among both elderly people and younger adults. In 1986 the prevalence increased from 4.8 per cent among those aged 30–39 years to 11.6 per cent among those aged 60–64 years; but through the next two five-year age groups and to 75+ years the trend was irregular, even falling from 13.7 per cent at 70–74 years to 12.5 per cent in the oldest category. Self-reported health surveys are influenced by age-specific expectations or norms as well as other contextual factors. In Singapore, standardized death rates in late life are falling but hospital admission statistics suggest that the health of elderly people is getting poorer. It is by no means easy to put together a convincing overall picture.

An accessible and particularly thorough study of elderly people's health was conducted in Fiji, the Republic of Korea, Malaysia and the Philippines on behalf of the World Health Organization (Andrews et al. 1986). This went to great pains to use validated and reliable questions and to minimize the distortions of translating them into different languages and technological and cultural settings. The study none the less found it difficult to synthesize the health situation of elderly people, not least because it was found that subjective assessments of health tend to be culture-specific. In general, the respondents' evaluations of their own health status revealed no age trends or sex differences. More objective measures, particularly on the prevalence of visual, dental and hearing problems and difficulty in walking, consistently showed that disability increased with age. The proportion of people who experienced difficulties with physical activities of daily living (eating, dressing, walking, getting in and out of bed, bathing and using the lavatory) was uniformly low. Only among those aged 80+ years and in Korea did the proportion exceed 1 in 10. As in developed countries, strong regional, urban–rural, income, social and caste differentials in health and mortality were found among the elderly populations.

AGE STRUCTURE, DEPENDENCY RATIOS AND 'MATERIAL SUPPORT'

Decreasing birth rates and the consequential rise in the elderly share of the population do not necessarily change the social situation of each elderly person, but they invariably alter the position of the age group in society. A common but superficial indicator is an age-structure ratio, such as the number aged 65 years or more in relation to those aged 15–64 years. In the less developed world, the number of people aged 65+ years for every 100 of those aged 15–64 years has increased from 6.9 in 1960 to 7.3 in 1990. By 2020 it is expected to increase to 10.6. While the change over the next thirty years represents an increase in the ratio of elderly people to younger adults of 45 per cent, from 1 to every 14 to 1 to every 9, the real change in dependency could be in the opposite direction. For example, there may be relatively fewer incapacitated elderly people in 2020 than now. What is less arguable is that young children invariably require support and divert their parents from 'productive' work. They require food, housing space, socialization and, increasingly, formal education. Their numbers relative to each 100 of the adult population increased from 65 in 1950 to 77 in 1970. Subsequently the decrease in the birth rate has brought this number down to 59, and it is expected to fall to 41 by 2020. Measured in these crude ratio terms, the overall ratio of children and elderly people to the adult population is likely to fall from 2 to every 3 adults (1.5) to under 1 to every 2 (1.9).

Age-structure ratios are often described as 'economic dependency' ratios. The assumption is that people over the age of 65 years are by and large unemployed or economically unproductive, whereas younger adults are productive. This assumption should be critically challenged, particularly (but not only) for low-income nations, for it is founded on an ignorance of social conditions. In most parts of the world, the predominant forms of employment – subsistence and peasant cultivation, domestic craft manufacturing, petty trading and personal services – after infancy are not confined to particular years of life. They are associated neither with formal retirement nor with pensions. People continue working as long as they are able to do so; they commonly have no choice, for they have no savings and no alternative source of material support.

In traditional settings, a person's economic contribution is often to a shared domestic economy, through work in the fields, domestic work or child care. Some elderly people will of course be more productive than others, because they are fitter, more energetic or living in situations which enable them to be productive, but the same can be said of younger adults. Age-structure, or economic dependency, ratios, can be even more misleading for societies where a larger part of the productive effort is not measured by monetary exchanges than in modern industrialized economies.

Westerners also need to interpret 'material support' with humility: it is inadequately conceptualized as money income. In less developed countries, particularly in rural areas, support has customarily been not in the form of wages or money payments, for either elderly people or younger adults, but more often through the exchange of services and goods. When the exchange is immediate and tangible, the transaction can be described as barter. A profound problem for the long-term situation of elderly people in rapidly modernizing countries is that exchanges are sometimes founded on the more complex obligations which are set up between family members over long periods. Parents nurture and raise their children in part from a customary expectation that later they, or their spouses or children, will provide reciprocal support. A credit of obligations in a 'social support bank' is established. As in Europe before this century, the majority of elderly people contribute to their domestic and local economies, if at a reduced level, through their seventh and eighth decades. When incapacity or illness makes this no longer possible, relatives, local communities and sometimes religious orders have provided a degree of support for most, although, as in the richest countries, cases of abandonment and total isolation do occur. In the industrialized world, this system has been profoundly modified if not completely replaced by the new social construction of old age and retirement.

The word 'development' implies positive change – it implies an increase in gross and per capita domestic product. But the experience of individuals may not show parallel change. For many elderly people in low-income countries, the predominant impression of development may be of unsettling change. The social and economic system in which they grew is disrupted. Customary expectations about their living arrangements, family situation and means of material support may be disappointed. The rapid development of recent decades has generally meant a rise in capitalist forms of production and the growth of waged and salaried employment. In comparison to traditional, peasant and domestic self-employment, the new jobs are inflexible as to hours of work, the tasks demanded and their intensity. If people become sick or weak, minor and incremental changes in what they do are incompatible with the regulated division and co-ordination of labour. The result is that redundancy is enforced and retirement is invented if not at first resourced. As a population becomes more dependent on money incomes, it becomes more vulnerable to lay-offs and enforced retirement. The need grows for public forms of income support and of domiciliary and residential services for elderly people. The most frail and dependent will inevitably be the sick and the socially isolated, and there is a tendency to place too much of the responsibility for the casualties upon the medical services. The mistake of Western countries, to over-medicalize the problems of frail elderly people, must not be repeated.

TRANSITIONS IN WELL-BEING AND HEALTH THROUGH DEVELOPMENT

It is a delusion to believe that governments or, as Auguste Comte hoped in the early nineteenth century, even positive social science and administration can to any great extent control the profound forces of economic and social change unleashed by rapid development. But they can be more than 'fire-fighters' who respond only to crises that cannot be ignored. A political force through development is for governments to involve themselves more and more in the health and welfare of their populations, as was the case with the sanitation and housing crises of the nineteenth-century British industrial cities. Governments and public authorities in low-income countries have very restricted resources. They will have to make agonizing choices, not always guided by popular democratic representative systems. There will be a pervasive impulse to promote 'development', often leading to a high priority for strengthening education and the health services for working-age parents and their children.

The real politics of health and welfare development therefore will tend to produce a consensus among politicians, trade unions, and health and welfare professionals that the interests of elderly people are both opposed to and of lesser priority than those of young families. The sequence of health and welfare innovations in industrializing Britain should be remembered. Compulsory elementary education was introduced in 1870. Following Charles Booth's demonstration of the concentration of poverty in both very young families and old people, and his lifelong campaign for a state pension, the first non-contributory scheme came in 1908. Geriatric medicine began to develop as a hospital speciality at about the same time; but in their early phase during the 1920s and 1930s, municipal and voluntary personal social services were almost exclusively concerned with vulnerable children. Public welfare measures for elderly people of more than rudimentary degree became widespread only during and after the Second World War.

The evolution of health and welfare institutions and services will of course differ in every country and era, but a general principle of the relation between development, health and well-being can be stated. This arises from the inevitable fact that as time passes, each person is successively an infant, a child, an adult and an old person. The principle is that, in the short term, the interests of different age groups are opposed, but in the long term they are identical or shared. Because the process of development plays out over a long period, the most profound inequities are arguably between successive birth cohorts more than between age groups. Development produces both individual gainers and losers, and disadvantaged and prospering cohorts. Those born at one period may be particularly harmed by disruptive changes, ranging from the attitudes and behaviour of their children to the collapse of their anticipated means of material support. A later cohort may, however, have adjusted to the new conditions and developed very low expectations, but then be happily surprised that their health, welfare and economic position improves. Birth and perceived fortune are interlinked. The practical agenda is not only to identify the age groups in the short term who are particularly harmed by the process of development, but also to try to understand the longer-term dynamics of the development process and its impacts on successive cohorts of the population.

PROSPECTS

The dominant response to demographic ageing in Washington, Westminster, Canberra and government treasuries elsewhere is a fiscal concern: that ageing is accompanied by a deteriorating ratio between taxpayers and social-security and health-service beneficiaries. This perception is a particularly distorting glass through which to examine the prospective circumstances of elderly people in most nations of the world. In many low-income nations demographic ageing has only recently begun and will probably be more rapid than in the developed world. The pace of the age-structure adjustment will most of all be governed by the progress of fertility. In itself, it has no necessary relation to the changing material, welfare and health circumstances of elderly people.

The situation of elderly people in many low-income countries will be largely a function of the nature of their socio-economic transformations. This includes the pace and comprehensiveness of urbanization, the 'reach' of capitalist and modern economic developments and the extent to which cultural norms as to the family and the roles of men and women are changing. Several contradictory trends can be identified. On the one hand, the spread of good-quality drinking water and sewerage systems suggests a reduction in ill-health and premature mortality in which infectious and parasitic pathogens are implicated. Also, progressive improvements in the standard of living and in modern food storage and distribution systems may slowly reduce the debilitation and vulnerability brought on by malnutrition. On the other hand, the adoption of some Western habits, notably cigarette smoking and refined drug use, may act against these improvements.

Of more general concern, but more intricate effect, is the disruption to traditional and anticipated family and household forms for elderly people. These have in the past provided access to material resources, health care, social roles, respect and emotional and psychological support. They are being weakened by urbanization, the spread of capitalist forms of employment, social mobility and attitudinal changes including the growth of careerist and materialist attitudes among younger people. The consequences for the housing standards of old people, and their nutrition, health and morale, will be intricate and difficult to understand. An optimistic view is that the frequency of catastrophic disruption will be low, but at the same time the accumulation among elderly people of affronts and disadvantage could well create very severe social and health problems early in the next century. The changing social formations among elderly people deserve very close study and exceptionally prescient powers.

REFERENCES

Andrews, G.R., Esterman, A.J., Braunack-Mayer, A.J. and Rungie, C.M. (1986) *Aging in the Western Pacific*, Manila: WHO Regional Office for the Western Pacific.

Chayovan, N., Knodel, J. and Siriboon, S. (1990) *Thailand's Elderly Population: A Demographic and Social Profile Based on Official Statistical Sources*, Research Report No. 90–2, Ann Arbor, MI: Population Studies Center, University of Michigan.

Keyfitz, N. and Flieger, W. (1990) *World Population Growth and Aging: Demographic Trends in the Late Twentieth Century*, Chicago: University of Chicago Press.

Kinsella, K. (1988) *Aging in the Third World*, International Population Reports P-95, No. 79, Washington, DC: United States Bureau of the Census.

Martin, L. (1988) 'The aging of Asia', *Journal of Gerontology* 43: S99–113.

Myers, G.C. (1985) 'Aging and worldwide population change', in R.H. Binstock and E. Shanas (eds) *Handbook of Aging and the Social Sciences*, New York: Van Nostrand Reinhold, pp. 173–98.

—— (1990) 'Demography of aging', in R.H. Binstock and L.K. George (eds) *Handbook of Aging and the Social Sciences*, 3rd edn, New York: Academic Press, pp. 19–44.

Phillips. D.R. (1990) *Health and Health Care in the Third World*, London: Longman.

—— (ed.) (1992) *Ageing in East and South-East Asia*, London: Edward Arnold.

United Nations (1985) *The World Aging Situation: Strategies and Policies*, New York: UN.

Warnes, A.M. (1992) 'Population ageing in Thailand: personal and service implications', in Phillips 1992, op. cit., pp. 185–206.

11

CARING FOR ELDERLY PEOPLE

Workforce issues and development questions

Alun E. Joseph and Anne Martin-Matthews

INTRODUCTION

Population ageing is a demographic process that has far-reaching implications. It affects ageing individuals and their families, communities and organizations charged with meeting the needs of seniors, and governments seeking to manage scarce resources. The debate about the social ramifications of population ageing almost always features a concern with economic or health care issues, often in combination. This chapter is concerned with an important sub-set of these issues, namely the workforce implications of care-giving to elderly people.

The notion of 'dependency', with its various connotations, is central to much of the debate on the social ramifications of population ageing. From an economic vantage-point, changes in the ratio of people in age groups typically not in the labour force (0–14 and 65+) to those typically in the labour force (15–64) are viewed in terms of their implications for public and private expenditure patterns. The consensus is that population ageing will result in a gradual transformation of the expanding dependent population, such that its centre of gravity will move from 'pediatrics' towards 'geriatrics' (Beaujot 1991). Subsequent economic implications are seen to stem as much, if not more, from this change in the nature of dependants as from the numbers involved. Foot (1989) has estimated that older Canadian dependants represent about 2.5 times the average cost of younger dependants in terms of public-sector expenditures, mainly because of pension and health care costs. Indeed, a number of commentators have observed that ageing of dependent populations represents a shift from private dependency costs (primarily borne by the family) to public costs. They have also noted that resource allocations for the young are very much an investment in the future labour force but that those for the elderly are almost totally taken up by immediate consumption.

Health and health care figure prominently in discussions of the costs of ageing, often with reference to the notion of dependency outlined above: 'In effect, the working population pays for improvements in health services that extend life, then it pays for the health and pension costs of persons who live longer' (Beaujot 1991: 226). We do not dispute this assertion, but contend that the focus on the direct costs of increased longevity, particularly in terms of hospital care, has tended to mask its more pervasive impacts. In essence, this institutional, public-sector emphasis ignores the fact that responsibility for the care of the elderly falls primarily to family members and not to the state. It has been estimated that in developed countries like Canada approximately 80 per cent of all assistance provided to the elderly comes from family members and from friends. Indeed, the formal care system is the least prevalent form of health care in society, and is typically accessed only after self-care and informal care by family and friends have been utilized. Thus, while the private costs of child care have invariably been entered in the dependency ledger, those for 'elder care' have not. This is all the more surprising because both elder care and child care are essentially workforce issues that can influence job performance by care-givers, and it is surely the viability of the workforce that determines the economic outcomes of increasing levels of dependency. Efficient workforces will be more able to support higher levels of dependency than will less efficient ones.

In this chapter we view elder care as a workforce issue with potentially profound implications for patterns of economic performance and development. We view this workforce perspective as particularly advantageous because it both encourages and permits us to acknowledge that the ramifications of population ageing are intertwined with those of other social trends. Specifically, we will explore the relationship between elder care and the evolving role of women in the workforce.

Notwithstanding the focus on Canadian data in the body of the chapter, our concluding discussion seeks to extend the North American experience to other countries. In particular, we are interested in the relevance of the elder-care–workforce issue in developing countries. We argue that the dynamics of modernization will affect the evolution of the elder-care–workforce issue in several important ways. Modernization will lay down the demographic parameters governing the potential relationship between workers and their elderly relatives. It will also influence norms of filial obligation and attitudes towards the involvement of women in the workforce.

POPULATION AGEING AND WORKFORCE TRENDS IN CANADA

Canada is now in that part of the demographic transition from high to low vital rates in which population ageing is driven by a combination of fertility decline and mortality reduction (Beaujot 1991). What makes the Canadian situation especially interesting from a demographic perspective is the country's positioning in the statistical vanguard of the 'baby boom'. The demographic imprint of this sudden and dramatic upsurge in the birth rate, lasting from 1946 to 1966 (Stone and Fletcher 1988), will give rise to extremes in ageing trends over the coming decades. Thus, while the proportion of the Canadian population aged 65 or older is projected to increase from 10.6 per cent in 1986 to 25 per cent in 2036 (Beaujot 1991), the ageing trend will not be a simple, linear one. Growth in the number of elderly people, currently standing at about 3 per cent per annum, will slow down in the late 1990s and early 2000s as the 'deficit cohorts' of the 1930s reach old age, but will rise again dramatically as the 'shock-wave' of the baby-boom generation attains age 65 (Stone and Fletcher 1988). Similarly, the parameters of the baby-boom phenomenon will affect the 'ageing of the aged', a particularly important sub-trend within population ageing. The population aged 65 or older is currently growing at a rate three times higher than that for the entire population, whereas the number of people in their eighties and nineties is growing four times faster. The baby-boomers will temporarily reverse this ageing-of-the-aged trend in the decade or so after 2011, but they will push it to even higher levels as they begin to reach age 80 in 2025.

A decline in the size of the labour force will roughly parallel the ageing trend to 2036 as the source population (aged 15–64) becomes smaller. The long-term trend is clearly towards increasing numbers of elderly dependants, such that by 2036 it is estimated that there will be one retirement-aged person for every 2.5 people in the labour force age group (Beaujot 1991). However, as noted earlier, a focus on dependency ratios can divert attention away from important changes in the workforce, such as female participation rates, that have ramifications for the way in which the structural potential for elderly dependency is translated into real economic impacts.

In 1951 only 23.5 per cent of Canadian women worked for pay outside the home, but by 1987 this had risen to 56.2 per cent, and a workforce participation rate in excess of 66 per cent (or two out of every three adult, non-elderly women) is expected by 1995. Increases have been particularly dramatic in recent decades, such that the number of women working outside the home grew by 42 per cent between 1976 and 1986. In comparison, the number of men in the paid labour force grew by only 12 per cent during the same period. As a consequence, women's share of total employment increased from 37 per cent in 1976 to 43 per cent in 1986 (Duchesne 1987). Canada now ranks behind only Scandinavia in its reliance on women in the workforce.

Regardless of these advances, the role of women in the workforce remains distinctive, especially in terms of the high incidence of part-time employment and the concentration of women workers in particular occupations and in a few industrial sectors. In 1990, 33 per cent of all employed women

in Canada were working part-time, with fully a quarter of them reporting that they did so involuntarily in the absence of full-time employment opportunities. This high incidence of part-time employment is
undoubtedly related to the relative concentration of women in the service
sector (at about 85 per cent in 1990, as opposed to about 60 per cent for
men), elements of which rely disproportionately on low-paid, part-time
employees. At a societal level, the reduction in the numbers of entry-level
workers in the coming decade will probably force an increasing reliance
on older workers, thus prompting more concentrated effort to retain
them.

THE CONTEXT OF THE ELDER-CARE ISSUE

In terms of the elder-care issues under consideration in this chapter, it is
important to cast workforce trends and employment patterns in the context
of changes in the 'family', broadly conceived to embrace a range of social
relations. Of primary importance is the reinvention of the family itself. The
model nucleated family of the 1950s – dad working, mum at home with the
kids – has lost its relevance for the majority of Canadians. Dual-earner
families now constitute a majority (62 per cent) of all husband–wife families,
and families now include single-parent households headed by divorced or
never-married adults (MacBride-King 1990). Some of these changes, notably
the growing plurality of dual-earner families, are clearly a consequence of
increased female participation in the workforce. As such, they are part of
the structural precondition for the elder-care–workforce issue. Others, like
the growth in the number of single-parent households, represent both an
imperative for workforce participation and a deterioration in the family
context of care-giving. It is the rapidity of these interrelated, female-focused
changes in the workforce *and* the family that has made elder care an
economic issue of contemporary importance. In this sense, the development
of the issue is outrunning the ageing of the population, which is occurring
at a slower rate.

Women, who provide the great majority of care to both the young and
the old, will increasingly be called upon to balance work responsibilities and
family commitments to the aged. Virtually all women working outside the
home will be asked to balance work and family life, but the challenge will
be felt particularly by certain groups. In terms of elder care, the sub-group
most frequently identified in the literature as 'at risk' is the 'sandwich
generation', members of which have children at home *and* elderly parents in
need of assistance (Brody 1990). The sandwich generation is typically
characterized in terms of workers from the age group 40 to 60 (and
especially those between 45 and 54). This same age group of women is
becoming an increasingly significant portion of the workforce as the baby-
boom generation ages. Although the sandwich-generation is a structural

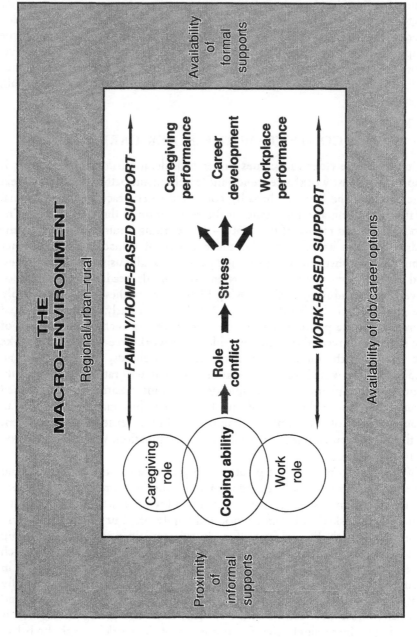

Figure 11.1 A conceptualization of the workforce implications of elder care

concept that carries no a priori expectations in terms of actual care-giving, it nevertheless represents the intersection of workforce trends, population ageing and care-giving patterns.

NATURE AND IMPACTS OF ELDER-CARE ISSUES

Figure 11.1 presents a generalized view of the ways in which the elder-care issue may translate into measurable economic impacts. It is important to note that the flow from overlap between work and family roles to job impacts is not seen as wholly deterministic. The psychology literature suggests, for instance, that some individuals cope successfully with overlap between work and family roles, and that stress does not always translate into measurable impacts. Indeed, involvement in multiple roles may actually be a coping device. Individuals may, for example, view their work responsibilities as respite from their care-giving responsibilities, and vice versa (Brody 1990).

Figure 11.1 distinguishes between the care-giving and employment impacts of a breakdown in work–family balance. Although not considered here, it is probable that poor performance in care-giving translates into additional demands on the formal health care system. Employment impacts are subdivided into workplace performance and career development components, distinguishing between short-term and long-term effects. Workplace performance impacts may include absenteeism and inappropriate use of sick days and vacation days, while career development impacts may manifest themselves in the refusal of promotions or job transfers. Such impacts obviously have direct economic implications in terms of the short- and long-term efficiency of the workforce, but they may also affect the mental and physical health of workers.

In order to test an expanded version of Figure 11.1, CARNET (the Canadian Aging Research Network) is currently engaged in a study of some 10,000 Canadian employees. Preliminary results of that analysis will now be presented, and we will begin by considering how elder care itself is conceptualized and measured.

Care-giving and care-givers

Any analysis of the relationship between work and elder care must begin with the recognition that 'there is no consensus among researchers, policymakers, service providers, caregivers, or care recipients themselves as to what constitutes family caregiving. And the definition of family caregiver varies widely' (Stone 1991: 724). Care-giving may be defined diversely by the types of care provided and its volume, intensity and duration, the relationship of the care-giver to the care recipient, and selected characteristics of the care receiver (including age, level of disability and presence of

Table 11.1 Percentage of males and females within each care-giving
category and their average age

Category	Females (N = 3,300)	Males (N = 1,793)
No care-giving	28.0% Age = 45 years	21.6% Age = 47 years
Have children under 18	29.5% Age = 40 years	32.2% Age = 42 years
Personal elder care (e.g. help with dressing, undressing, taking medication, feeding, etc.)	7.7% Age = 46 years	4.3% Age = 50 years
Instrumental elder care (e.g. helping with laundry, transportation, finances, arranging services, etc.)	6.0% Age = 40 years	5.4% Age = 40 years
Have children under 18 and personal elder care	5.0% Age = 42 years	3.6% Age = 45 years
Have children under 18 and instrumental elder care	23.9% Age = 43 years	33.0% Age = 45 years

Source: 1992 CARNET data

certain health conditions). Some studies define an individual as a care-giver
only if (s)he provides assistance in one or more activities of daily living (such
as bathing or toileting). In other cases, care-givers are much more broadly
defined to include those providing financial or emotional support to elderly
relatives. Yet other studies, among them several of the better-known United
States investigations of work and elder care, rely solely on a self-definition
of oneself as a care-giver (Scharlach *et al.* 1991).

In order to attempt a more precise definition of what actually constitutes
elder care, our research has anchored the measure of care-giving behaviour-
ally. We asked Canadian employees to indicate, across a broad range of
helping situations, the frequency with which they helped their elderly
relative during the previous six months. Helping situations ranged from
assistance with the activities of daily living, through help with household
chores, shopping and transportation, to financial assistance. Responses
ranged from never, through once a month or less, through once a week, to
daily.

Table 11.1 illustrates the range of care-giving responsibilities carried by
the employees participating in our study. About 25 per cent of the
respondents have no care-giving responsibilities, while a further 31 per cent
have responsibility for children only. In contrast, fully 44 per cent of all the
employees participating in our study provided some kind of assistance to
an elderly relative in the six months prior to the survey. When we consider
those who reported elder-care responsibilities *exclusively* (that is, they did

not also have dependent children), we find that 14 per cent of the women and 10 per cent of the men have such responsibilities. Many employees, however, have dual responsibilities, both to children under age 18 and to elderly family members. This is true of just over a third (37 per cent) of the men and over a quarter (29 per cent) of the women we studied.

In order to achieve greater specificity in studying the relationship between elder care and labour force participation, we categorized the nature of the tasks performed to yield two measures of elder care. One measure includes the demanding personal care activities associated with bathing, toileting, feeding, dressing and assisting with medication. We call these tasks 'personal elder care'. In addition to these activities, those providing personal elder care are invariably assisting with other tasks. The employees whose care-giving activities exclusively involve the provision of personal elder care constitute 8 per cent of the women we surveyed and 4 per cent of the men. By way of comparison, we also measured elder care in terms of the provision of assistance with more instrumental activities of daily living, such as assisting with transportation or finances and arranging services. Five per cent of the male employees and 6 per cent of the female employees we surveyed were providing this kind of 'instrumental elder care' exclusively.

We also wanted to tap the prevalence and characteristics of that group of employees who are emblazoned in the public consciousness as the 'sandwich generation'. The data in Table 11.1 indicate tht 5 per cent of the employed women we studied have children under age 18 and are also providing personal elder care. Four per cent of the men do so. This is the group that one might speculate would experience the most difficulty in balancing work and family responsibilities.

Role conflict and stress amongst care-givers

In order to facilitate discussion of role conflict and stress among care-givers, we will compare four groups: those with no care-giving responsibilities; elder care only; child care only; and dual responsibilities. Overall, care-givers with dual responsibilities were collectively more likely to experience difficulties in balancing work and family life. On a measure of perceived stress, the employees in the CARNET study who were providing care, be it to children or elderly relatives, reported significantly higher levels of stress than those with no care-giving responsibilities. Notably, employees who provide *personal* elder care and who also have responsibilities for children under age 18 reported the highest levels of stress.

In overall measures of work–family overlap, there were again significant differences between workers with care-giving responsibilities and those without. Those most likely to feel that their work responsibilities interfere with family life are those with dual responsibilities for both children and elders. In a parallel measure of the extent to which family is perceived as intefering with work, employees with dual responsibilities are again most likely to express difficulty in balancing work and family. In both cases, it is the employees who have responsibilities for children under 18 and who also are providing *personal* elder care who report the greatest difficulty in balancing their work and family lives. This is the same group of employees who report the highest levels of stress in their lives.

Employment effects

The final element in the progression set out in Figure 11.1 relates to the nature of career/work impacts. The data in Table 11.2 illustrate the range of job effects reported by employees in the CARNET survey. These effects, all occurring within the preceding six months, were reported to be as a result of responsibilities and activities outside work. In virtually every instance, significantly more employees with dual responsibilities for a child under age 18 and for *personal* elder care report job effects than do those in other care-giving situations. These effects include being late for work, taking sick days when not sick, being absent from work for three or more consecutive days, missing meetings and business trips and not seeking a promotion for which one was eligible.

The CARNET data show that employees with personal elder-care responsibilities had, in the six months prior to the survey, taken the most sick days when they were not sick in order to balance their work and family responsibilities. Those whose sole care-giving responsibility was for personal elder care averaged 5.4 such days, compared to 2.2 days among those with children under age 18, and 2.3 days among those with no care-giving responsibilities. It is very important to acknowledge gender differences in these workplace experiences; for instance, 26 per cent of female respondents with dual responsibilities for a child and for personal elder care reported missing meetings for family reasons, while only 14 per cent of the men in a similar care-giving situation did so. Yet more evidence of the career impacts of coping with work and family, especially among women, is furnished by the fact that fully 24 per cent of the women (but only 4 per cent of the men) with dual responsibilities involving personal elder care indicated that they had not sought a job promotion for which they were eligible.

It is appropriate to round out this discussion by considering absenteeism, a workplace phenomenon often attributed to a breakdown of coping ability. The economic importance of absenteeism is reflected in estimates that it typically costs employers 1.75 times an employee's wage rate, and that such costs could typically run to 8 per cent of the annual pay-roll (MacBride-King 1990). The CARNET data (Table 11.2) indicate that employees with personal elder care (alone or in combination with child care) are markedly more likely to exhibit this critical job effect than are other care-giver groups.

Table 11.2 Job effects associated with balancing work and family responsibilities among employees

Care-giver group	Reduced hours worked per week (%)	Late for work (%)	Take sick days when not sick (%)	Absent from work for 3+ days (%)	Use vacation time for responsibilities (%)	Missed meetings (%)	Missed business trips (%)	Did not seek promotion (%)
No care-giving	2	7	5	2	17	5	2	4
Child under 18	3	12	11	3	28	12	5	13
Personal elder care only	5	13	8	7	40	17	5	12
Instrumental elder care only	1	9	8	2	26	8	5	7
Child under 18 and personal elder care	3	13	15	7	47	23	16	18
Child under 18 and instrumental elder care	3	11	10	3	30	12	5	12

Source: 1992 CARNET Work and Eldercare Research Group data

DISCUSSION

The CARNET data confirm the existence of elder care as a workforce issue. As in other studies in Canada (MacBride-King 1990) and in the United States (Scharlach *et al.* 1991), it is women workers who apparently face the greatest test in terms of balancing work and family responsibilities. Studies of elder care are not yet advanced enough to yield unequivocal estimates of the economic costs embodied in job effects such as declined promotions and absenteeism. However, it is clear that such costs do exist, and that their magnitude will increase as communities and workforces age. It is to this 'future of elder care' that we now turn. Although our initial emphasis is again on Canada, we quickly turn to a broader theme, namely the relationship between the elder-care issue and economic development.

The future of the elder-care issue in Canada

As noted earlier, the immediate future of the elder-care issue in Canada is bound up as much with interrelated trends in female workforce participation and family structure as it is with population ageing. By the year 2000 high participation rates of women in the paid labour force and the ageing of the baby-boom generation will together move elder-care issues towards the forefront of the work–family agenda. What will remain constant as these trends work themselves out is the special vulnerability of care-givers who provide extensive support to both young and old dependants.

The future contours of elder care as a workforce issue may well depend on the answer to a simple question – whose issue is it? There is persuasive evidence that the majority contemporary opinion (among workers, employers and governments) is that elder care is very much a private issue. However, this attitude denies the cost of inhibited workplace performance or sub-optimal career development, both for employers and for society at large. In an increasingly competitive global economy, losses in worker performance and the systematic underutilization of human capital may become unacceptable. In this way, the interests of workers, employers and governments could well coalesce to promote a more proactive stance concerning support for workers experiencing difficulty in balancing work and family responsibilities. Indeed, there is evidence in Canada and the United States that progressive employers are already committed to a more aggressive stance on the elder-care issue (Scharlach *et al.* 1991). Even from this limited experience it is apparent that initiatives/interventions in the area of elder care will be much more diverse than those evident in the past for child care. Initiatives directed towards child care typically address needs that are both of more predictable duration and more homogeneous. Those directed towards elder care will likely have to be more idiosyncratic and of unpredictable duration. Some interventions will address the health care needs of elderly dependants, while others will attempt to promote the mental and physical health of care-giving workers.

Beyond Canada: elder care and economic development

It is always difficult to embark on the cross-cultural transfer of complex issues. Nevertheless, it is necessary to identify at least the key parameters that underlie the generality of the elder care–workforce issue. In countries like Australia, New Zealand and the United States, which have baby-boom generations comparable in relative size to that of Canada and which display similar trends in the role of women in the labour force, the elder-care issue will unfold in a very similar fashion. It is also apparent that elder care is becoming a workplace issue in demographically more mature countries in Western Europe, such as the United Kingdom, where there are high levels of female participation in the workforce. What is less clear is the relevance of the Canadian experience to developing countries. However, it is very likely that the evolution of the elder care–workforce issue in developing countries will be bound up with the dynamics of modernization.

Our central thesis is that modernization, which we take to be a euphemism for rapid economic and social change, touches on the elder care–workforce issue in several ways. In this discussion, we emphasize the broad demographic parameters of modernization that have implications for elder care. We also consider the ways in which these demographic trends may interact with cultural variables, particularly those governing the involvement of women in the workforce. We will focus our discussion on East and South-east Asia. This region presents a range of cultural contexts and development 'stories', including the newly industrialized countries (NICs) of Singapore, Taiwan, Hong Kong and South Korea, along with major demographic powers like China and Indonesia (Phillips 1992).

There is almost universal agreement concerning the strong association between modernization and the demographic transition to smaller families and an ageing population, although consensus on the direction of cause and effect is more elusive. The transition to a norm of smaller families, often accomplished over a relatively short period of time, has important ramifications for the 'arithmetic' of intergenerational linkages between workers and their elderly relatives. Put simply, there will be proportionately fewer young(er) family members to look after elderly relatives surviving to progressively older ages. This structural effect is most apparent in China, where it is the legacy of aggressive family planning policies. In that country, there will be 90 million or so persons aged 65 or over by the year 2000, and an emergent 4–2–1 population structure: 4 grandparents, 2 parents and 1 child (Phillips 1992). The potential strain on young workers is obvious, but actual economic impacts may well be determined by cultural effects.

Cultural factors affect modernization in various ways, of that there is no

179

doubt. Here, we will explore two cultural variables: the role of the family in caring for the elderly and the involvement of women in the workforce. Each relates clearly to the elder care–workforce issue. Turning first to the family, it is clear that in contemporary East and South-east Asia, it 'is being regarded almost universally as the principal supporter, carer and home provider for elderly people and particularly those unable to look after themselves' (Phillips 1992: 15). There will undoubtedly be strong resistance to the reduction of filial obligation in many countries, especially those espousing orthodox Muslim values. However, across the region one of the outcomes of smaller family size and increasing numbers of elderly may well be an increased demand for formal housing and support services. Indeed, it is instructive to note that such demands are already beginning to emerge in countries like Hong Kong and Singapore (Phillips 1992). Even if formal support does not reach the 20 per cent plateau common in many Western countries, the low level of current provision means that increases in rates of housing and service provision may be very dramatic. It is obvious from our earlier discussion, though, that the mix between formal and informal care for the elderly, *and* its workforce implications, will depend on the status of women.

The critical factor affecting the unfolding of the elder care–workforce issue in developing countries may well reside in the values that are placed on the direct (workforce) and indirect (family care-giving) contributions of women to society. There is only limited workforce-related evidence on this issue in the NICs, and even less in most other countries in the region. Fortunately, Japan, with its longer history of modernization, provides some illustration of the stresses involved in coping with work–family balance against the backdrop of pervasive demographic and social change.

Although women still constitute a smaller proportion of the workforce in Japan (37 per cent) than they do in Canada (45 per cent), recent increases in the female labour force participation rate have been very similar. Momentum for the increased involvement of women in the workforce has undoubtedly come from the chronic shortage of labour in Japan, but it has been facilitated by changing attitudes towards working mothers, and sanctioned by legislation (enacted in 1985) that prohibits discrimination in employment on the basis of gender (Tanaka 1990). Nevertheless, there still remains a strong cultural norm in Japan which dictates that married women owe their primary responsibility to their family life rather than to their careers. Although this ambivalence towards the role of women in the workforce – where should their priorities really lie? – undoubtedly exists in most cultures, it clearly has more force in some. In countries like Japan (and perhaps in the NICs too) the loss of female careers and the incidence of lower work performance because of care-giving pressures are made more palatable by the inferior position of women in the workforce. Since care-giving is dominated by women, the future of elder care as an economic (and political) issue may depend quite directly on the enhancement of women's role in the workforce. If the loss of productivity traceable to elder care is shown to be greater than the cost of supplying support to workers (at home or in the workplace), the issue is likely to attract policy initiatives.

In closing this discussion, it is appropriate to return to our contention that increased longevity (an integral part of population ageing) is a health-related phenomenon with profound social and economic ramifications. Although the fact of more older people living longer has obvious implications for the cost of health care and social welfare programmes, many of the long-term impacts of population ageing will be played out in the workforce arena. As long as elder care remains primarily a women's issue, its implications for economic performance and social productivity will be intimately bound up with the unfolding role of women in the workforce.

ACKNOWLEDGEMENTS

We are grateful to Bonnie Dunnett and Jo-Ann Hutchison, who ably provided research assistance. The financial support of CARNET (the Canadian Aging Research Network), funded through the Government of Canada's Networks of Centres of Excellence Program, is also acknowledged, as is the intellectual contribution of its various members.

REFERENCES

Beaujot, R. (1991) *Population Change in Canada: The Challenges of Policy Adaptation*, Toronto: McClelland & Stewart.

Brody, E.M. (1990) *Women in the Middle: Their Parent-Care Years*, New York: Springer.

Duchesne, D. (1987) 'Annual review of labour force trends', *Social Trends* (Canada): 26–32.

Foot, D.K. (1989) 'Public expenditures, population ageing and economic dependency in Canada, 1921–2021', *Population Policy and Research Review* 8: 97–117.

MacBride-King, J.L. (1990) *Work and Family: Employment Challenge in the 90s*, Ottawa: Conference Board of Canada.

Phillips, D.R. (ed.) (1992) *Ageing in East and South-East Asia*, London: Edward Arnold.

Scharlach, A.E., Lowe, B.F. and Schneider, E.L., (1991) *Elder Care and the Work Force: Blueprint for Action*, Lexington, MA: Lexington Books.

Stone, L.O. and Fletcher, S. (1988) 'Demographic variations in North America', in E. Rathbone-McCuan and B. Havens (eds) *North American Elders*, New York: Greenwood Press, pp.9–33.

Stone, R. (1991) 'Defining family caregivers of the elderly: implications for research and public policy', *The Gerontologist* 31(6): 724–5.

Tanaka, Y. (1990) 'Women's growing role in contemporary Japan', *International Journal of Psychology* 25: 751–65.

12

PRIMARY HEALTH CARE AND SELECTIVE PHC
Community participation in health and development
Sheena Asthana

INTRODUCTION

In 1978 at the joint World Health Organization (WHO) and United Nations Children's Fund (UNICEF) conference in Alma Ata, the principles of primary health care (PHC) were formally established. The approach has since been endorsed as the strategy to achieve 'Health for all by the year 2000'. The recommendations of the Alma Ata conference were seen as a breakthrough in official policy formulation, for they stemmed from an explicit recognition that the promotion of health depends upon improving socio-economic conditions and the alleviation of poverty. By highlighting the environmental, social and economic determinants of health status, the Alma Ata declaration recognized that health could not be attained by the health sector alone. The strategy called for an intersectoral approach that addressed the broader issue of underdevelopment.[1]

With regard to the provision of health services, the declaration argued that the centralized, technological health care model of the developed world was inappropriate to the needs of developing countries. Not only were modern health care facilities geographically and financially inaccessible to the majority of Third World populations; their curative focus was of dubious benefit in countries where the majority of diseases were preventable. The solution, it was argued, was to achieve universal coverage of basic health services by extending health centres and dispensaries, training paramedical health workers, using appropriate health technology and focusing on prevention.

The architects of PHC drew inspiration from China, Cuba, Tanzania and Kerala State in India in order to demonstrate that high levels of health and social development could be achieved by a political commitment to policies based not on economic growth but on social justice and equity. The Chinese health system attracted attention for its apparent focus on equitable socio-economic development, health service decentralization, preventive measures and mass participation. Particular attention was paid to the Chinese 'bare-foot doctors', who not only extended basic health care to rural communities but were theoretically accountable to the populations they served. Community participation came to be seen as a central component of PHC: not only would the active involvement of individuals and communities ensure that health programmes developed in response to locally felt needs; participation was also a way of raising political consciousness and of encouraging people to get involved in the wider development process.

The PHC strategy endorsed at Alma Ata was given almost universal rhetorical support. In practice, however, governments and international agencies have rarely pursued the goal of comprehensive change in the sense of altering the social, economic and political structures of unequal societies. Instead, intersectoral action has come to mean an integrated package of programmes implemented at the level of individual settlements. Nor has the commitment to a more equitable distribution of health services been realized. The PHC approach was launched at a time of deepening recession and growing Third World debt. The 1980s saw the widespread implementation of economic adjustment policies aimed at transforming less developed countries into more market-oriented economies (see Chapter 3). Forced to make cuts in their public expenditure, many Third World governments have allocated scarce resources to the tertiary health sector at the expense of primary levels of care.

The new economic and political climate thus led health policy-makers to reassess the comprehensive PHC strategy. Many concluded that it was too idealistic and that, given financial and practical constraints, governments would be better advised to select and target diseases on the basis of prevalence, morbidity, mortality and feasibility of control (including efficacy and cost). This new approach, which is defined as selective primary health care (SPHC), places particular emphasis on paediatric conditions, high-priority diseases including diarrhoeal diseases, measles, whooping cough and neo-natal tetanus (Walsh and Warren 1979). It was given substantial support in 1982 when UNICEF identified four specific 'social and scientific advances' for improving child health and nutrition.[2] Low-cost interventions such as oral rehydration therapy and immunization proved particularly attractive to donor agencies as they produced 'results' and appeared to optimize the cost-effectiveness of primary health care programmes. Consequently, since the early 1980s national policies have tended towards more selective, almost vertically organized, programmes rather than integrated community-based health programmes requiring intersectoral co-ordination.

SPHC has given rise to considerable controversy (Rifkin and Walt 1986). Its critics suggest that it deals with diseases in isolation rather than tackling the root causes of ill-health. As child mortality in less developed countries is often due to a combination of malnutrition and repeated bouts of different infections, children who are protected against a very few diseases may simply die from other causes. There is a moral dilemma, too, about whether priority should be given to diseases that cause high child mortality. SPHC ignores the many chronic, non-communicable complaints that affect adults and which can compromise their ability to support their dependants (Unger and Killingsworth 1986).

Perhaps the most fundamental criticism of SPHC is that it represents a return to a technologically and cost-oriented approach which denies local initiative. The identification of priority diseases and the implementation of specific technologies designed to reduce those diseases naturally imply top-down intervention. By limiting the definition of community participation to that of a means of improving the acceptability of services and the efficiency of service delivery, selective intervention undermines the promotion of self-help and the potential of local organizations to play an active and direct role in the struggle for health. For those who support comprehensive PHC and who place community participation at the heart of the health and development process, SPHC is seen as 'counter-revolutionary'.

The rest of the chapter explores the different approaches to community participation that are expressed in the debate on PHC and SPHC. Having outlined the arguments for participation and identified two opposing trends in its evolution in practice, some of the problems faced in promoting participation will be considered.

OBJECTIVES OF PARTICIPATION: MEANS OR END?

Objectives of participation range from economic and practical concerns associated with project efficiency, relevance and cost recovery to political aims of equality and empowerment. The nature and scope of participant involvement vary accordingly. One theme is that, as the rural and urban poor know their priority needs better than any outsider, community involvement in problem selection and programme planning ensures a greater relevance and commitment to the project. Another argument is that, if people participate in the execution of projects by contributing their knowledge, skills and other untapped resources, governments can assist a far greater number than can be reached by conventional programmes. Resource contributions can take a variety of forms, of which cash payments are the most controversial. Whereas proponents argue that community financing may be the only way of overcoming the lack of funds for primary health care, opponents argue that it places the burden of financing health care on the people least able to afford it.

In the definitions outlined above, participation is seen primarily as a *means* to get something done, whether this be the enhancement of project effectiveness, project efficiency, the sharing of costs or a combination of these goals. Such predetermined objectives imply the presence of 'top-down' planning. At best, participants will be enlisted to make or shape decisions about what specific project activities should be implemented in a programme, how activities will be undertaken and who will be involved in the various stages of implementation. Their capacity for future participation may also be enhanced. A more limited conception views participation as the sharing of project benefits. Such involvement is better referred to as 'utilization'.

Although most proponents of participation as a 'means' do not couch their proposals in explicitly political terms, a common criticism of this approach is that, while it differs from 'modernization' strategies by recognizing that the benefits of economic growth at the national level do not necessarily 'trickle down' to the poor, by implementing programmes within existing national and international economic orders, the approach fails to confront the structural causes of poverty. For some critics it amounts to little more than social cushioning: poverty may be temporarily relieved but it will not be eradicated.

An alternative view to that which perceives participation as an instrumental means is that which regards participation as *an end* in itself. This perspective sees participation as a dynamic and unpredictable *process* which should arise from the grassroots rather than being imposed from above (Oakley 1989). Although this approach is compatible with Western democratic ideals (which state that ordinary citizens have a right to share in decision-making), it is normally associated with radical theorists who argue that participation is a process of *empowerment* of the deprived and excluded. This view is based on the recognition of differences in political and economic power among different groups and classes. The critical issue is that oppressed groups, who lack power because of their class, sex, caste, ethnicity, and so on, break their dependence on groups that exploit them. Participation thus implies a struggle for the redistribution of power and resources in society.

It is clear that different interpretations of participation are very much subject to political ideology. *Means* definitions rely on a *consensus* theory of society, in which the state embodies the interests of society as a whole and is interested in ensuring that poor communities derive real benefits from national and local development efforts. *Ends* definitions are more closely linked to *conflict* accounts of society in which different interest groups struggle for control of available assets and resources. As the state is seen to act on behalf of the ruling classes, furthering their interests, the accumulation of wealth and the concentration of power, the empowerment of the poor is the only way in which they can meaningfully benefit from the fruits of development.

ANTECEDENTS OF THE COMMUNITY PARTICIPATION APPROACH

Despite the popularity of community participation in PHC literature, the health sector was a relative latecomer to the concept. Much of the rhetoric that characterized the community development (CD) movement of the 1950s and 1960s can be seen in project documents today. CD programmes comprised a wide range of development activities, including agricultural improvements, water and sanitation, infant and maternity welfare and mass education. Emphasis was placed on identifying and planning for locally felt needs; on building up local initiative through training local leaders and encouraging self-help efforts; on co-ordinating technical assistance in order to bring about the integrated development of local communities; and on using multi-purpose village-level workers to motivate people to improve their own living conditions.

Although the aim of self-help through the development of communities appears to be similar to current ideas about participation, radical theorists are vociferous critics of CD. They suggest that in practice it more closely resembled a bureaucracy than a development strategy, with superimposed direction and limited participation in goal setting and programme formation. A second problem was that CD assumed that communities were homogeneous and had the same values and aspirations as the national government. By ignoring existing socio-economic structures and allowing benefits to be captured by powerful local groups, CD programmes perpetuated structures of inequality at both the local and national levels.

In contrast to the CD movement, the concept of conscientization rests upon the belief that societies are based on conflict and inequality. This movement is associated with the Brazilian educator Paolo Freire, who argued that the key to development lies in creating a critical awareness among the poor of the real causes of their oppression and of their capacity to change their situation. Under this approach, which is closer in sentiment to the radical definition of participation as an end in itself, the role of community-level workers is essentially political. However, conscientization is subject to a number of criticisms. The assumption that dominated groups require the intervention of outside campaigners before finding out that they are exploited is paternalistic. The movement has also come under attack for disregarding the fact that, by challenging the existing order, conscientizing activities are likely to meet with confrontation.

Community development and conscientization offer quite different interpretations of participation. The former involves people in development programmes devised from the top-down; the latter empowers people to undertake development activities themselves. Yet the ideas and assumptions of *both* approaches were incorporated into the Alma Ata declaration on PHC. On the one hand, the call for equity and social justice implied a politics of empowerment. On the other, the goal of reforming service delivery systems led to a more limited definition of participation as a management tool (Hollnsteiner 1982). As a result, the role of community participation in primary health care has become very ambiguous. Policy rhetoric continues to emphasize the potentially liberating role of community involvement in health and development. In practice, however, participation is given a very limited role in project activities. In the next section, some of the problems faced in translating the rhetoric of participation into reality will be considered.

PARTICIPATION IN PRACTICE: CRITICAL ISSUES

The context of participation

As suggested above, inspiration for the participatory approach to PHC came principally from developing socialist countries. In the transition from national experiences to global strategy, insufficient attention was paid to the exportability of the participation approach. The health movement in China was especially idealized, accounts failing to recognize that much of the initiative for mass organization in this highly controlled society came from above. It was only later that questions began to be asked about whether the Chinese model could be replicated in different socio-political settings.

The political context

If local communities are to participate meaningfully in the design and implementation of development activities, there must be a political commitment at the national level to bureaucratic reorientation. In most parts of the world, however, there has been almost no decentralization of effective authority and power. Only in a few socialist countries has provision officially been made for decision-making at the neighbourhood level. The Ethiopian peasants' associations, for example, were empowered to redistribute expropriated lands, institutionalize a communal form of agriculture and encourage local infrastructural development. Like the Chinese communes and the Committees of the Defence of the Revolution of Cuba, the Ethiopian peasant associations have been admired for their mobilizing capacity. However, as all of these organizations are very much geared towards fulfilling the predetermined goals of the state, they are more often interpreted as vehicles for social control.

Although liberal regimes grant their citizens political representation, they do not necessarily offer greater opportunities for participatory development. Because people's empowerment implies a change to the political status quo, governments may profess to support participation while deliberately maintaining a narrow power base. By creating a bureaucratic framework for channelling local demands, for example, a government will be seen to be responsive to the needs of the poor. In reality, however, grassroots mobilization is contained and participation routinized. Local leaders may be co-opted through the practice of political patronage. Here an established or aspirant politician seeks political support from the poor. In return for votes, he will use his influence to secure local services. Community members remain dependent upon the goodwill of their patron and fail to realize their potential for autonomous action. Finally, governments may be under pressure to skew their policies towards donor-led interests. Today, many governments that previously advocated collective responsibility for health care are increasingly favouring the private market which offers little scope for community involvement. Within the public health sector, the goal of authentic participation is often undermined by the tendency to concentrate on specific *medical* targets rather than comprehensive social development.

The socio-cultural context

Community participation is promoted as a global strategy. However, attitudes towards collective action vary according to socio-cultural context. In many parts of Asia, for example, traditional authority relations continue to dominate village life, the tendency being to follow strong leadership rather than make collective decisions. Such concentrated leadership offers limited scope for participation in decision-making. However, village headmen are in a powerful position to mobilize community members to undertake pre-planned development activities.

The historical relationship between the church in Latin America and the landowning aristocracy provides the backdrop for a major conscientizing force that is peculiar to this region. Church activists inspired by liberation theology rejected the élitist, corrupt practices of their predecessors and were at the forefront of resistance movements against authoritarian regimes. Committed to promoting community-based action among the poor, they work with so-called *comunidades de base* (base communities) to promote co-operative solutions, self-help and participatory democracy. Particular emphasis is placed on consciousness-raising in order to mobilize the masses in support of a radical change in social and productive relations.

While cultural values may be used to promote collective consciousness, there is always the danger that religious, caste or clan divisions will have a negative effect on participation. In India, for example, where members of low castes are seen as polluting to higher-caste groups, promoting collective responsibility for facilities such as water taps or kitchen gardens can prove difficult. In rural Gambia, factionalism based on religious controversy and caste heterogeneity has been found to disrupt programmes where community participation was desired. In cultural contexts where women are seen as repositories of traditional and family values, there may be great resistance to women's participation. Because women's position in society and the way it affects the health of their children are an issue of clear concern to governments and international agencies, more attention should be paid to ideological barriers to female mobilization.

Table 12.1 A check-list for identifying participatory components of projects

Stages of the administrative process	Activity	Examples
Planning	Needs assessment	Role of community members in research and analysis of health needs; assessment of non-health needs; how are results used?
	Establishment of goals and priorities	Discussion of goals and priorities. Are all groups represented?
	Programming	Development of proposals; discussion of activities and collective action; flexibility of project to respond to new goals and priorities.
Implementation	Community-based health activities	Training of community health workers and traditional mid-wives; detection and monitoring of groups at risk; performance of home visits; provision of vaccinations; oral rehydration services; growth monitoring and nutrition programmes; health education; provision of community crèche.
	Community development	Water, sanitation and housing; schools and adult education programmes; kitchen gardens; income generation and credit schemes; community co-operatives; distribution of assets; consciousness-raising activities; programmes designed to raise women's security and status.
	Project utilization	Community mechanisms for the dissemination of information about project activities; use of formal and informal motivators; who are the target and actual beneficiaries?
	Resource mobilization	Beneficiary contributions (money, labour and materials) to project activities; how are contributions raised and are equal contributions expected from all members? Who decides how community resources will be used?

Table 12.1 Continued

Stages of the administrative process	Activity	Examples
Control	Management	Role of health committees and other community organizations in project management; how are leaders and office-bearers elected? Is leadership representative and accountable? What is the relationship between community organizations and project managers?
	Supervision	Lines of responsibility for project management; role and accountability of community health workers and other project staff.
Evaluation	Evaluation	One-off, periodic or continuous monitoring of progress? Role of community members in assessment; are general meetings held for the presentation and discussion of results?

Source: after Agudelo (1983)

Project characteristics

A wide range of project characteristics influence the nature and scope of participation in any given context (Table 12.1). While several agencies provide 'check-lists' of critical factors that influence the participative process, care should be taken in assuming that community involvement can be planned for. Nevertheless, it is commonly agreed that the objectives and content of the project, the nature of its implementing agency, the role and training of project staff, and the mechanisms for the promotion of participation all have a bearing upon what type of participation is engendered.

Objectives of the project

Health planners' interpretations of the role of community participation are closely linked to different definitions of community health. Rifkin (1985) distinguishes between three approaches. *The medical approach* sees health as the absence of disease, to be brought about by medical interventions based on modern science and technology. Community participation in this context is a *means* by which medical professionals can increase the efficiency of the

190

health services they deliver. Health programmes are planned and implemented within vertical, top-down structures, and a passive, limited role is envisaged for the community. One example of this approach is that of community-based distribution by lay personnel of contraceptives. Another is the use of community registers and local motivators to reach immunization targets. Because community involvement is restricted to specific activities or events, it tends to be temporary in duration.

The health planning approach, which characterizes PHC in practice, is based on the view that health is essentially the result of the appropriate delivery of health services. It recognizes that health care resources in many Third World countries are inequitably distributed and that the health care sector must be restructured to provide the most benefits to the greatest number of people. Both *end* and *means* objectives of participation are incorporated into this approach. Local people have a right to participate in activities that affect their lives and must be involved in all aspects of programme planning if they are to be committed to the project. At the same time, emphasis is placed on the practical need to mobilize community resources in order to extend service provision. Participation is seen, therefore, as a form of *collaboration* between the development agency and the community.

An example of the health planning approach in practice is the Kasa Mother–Child Health–Nutrition Project in Maharashtra State, India (Shah 1976). The project, which was initiated by the state government, was explained to the communities, who were asked to select local part-time social workers (PTSWs). These workers, who were to be trained and supervised by existing primary health care staff, were taught to keep a community register, to record child weights, monitor pregnant women, treat minor illnesses, educate women about nutrition, health and family planning, distribute supplementary nutrition, assist PHC staff in immunization programmes and chlorinate drinking-water wells every month. Community involvement continued to be sought in the programme, PTSWs meeting regularly with their villagers in order to inform them about 'at-risk' children and to discuss social problems and solutions. Despite some problems, for instance in recruiting eligible women to work as PTSWs, the project did manage to secure active participation from community members. Examples of community assistance include the provision of sites on villagers' verandas for the distribution of nutrition supplements and of milk to children whose mothers had lactation failure. Local members also helped to feed and accommodate the immunization teams, to round up children for their vaccinations and to construct roads to improve communications with the PHC staff.

The community development approach, which should not be confused with the earlier CD movement, believes that health improvements are not solely dependent upon direct health-sector activities, but can only be achieved through better living conditions. According to this approach, services are a tool to be used in the education of people towards their participation, and health is only one of a number of possible entry points into the development process. It stresses that programmes for participation must start with awareness-building and that project activities should be developed in response to felt needs and self-help. In contrast to the medical approach, which harnesses local efforts to specific short-term goals, the community development approach sees participation as a long-term process in which community confidence, solidarity, responsibility and autonomy are gradually built up.

Although this emphasis on 'bottom-up' development is very popular in the primary health care literature, there are very few development *projects* in which the community plays a dominant role. The best examples of *self-help* are drawn from a period when many government authorities were unwilling to support development activities and collective self-help was necessary for survival. In the well-known squatter settlement of Villa el Salvador in Lima, for example, the community organization provided food for groups of people during the initial stage of illegal occupancy; set up committees for schools and health care; and formed a community bank which provided credit for communally owned enterprises and arranged loans for the provision of water and electricity.

Since the 1970s, state involvement in basic service provision has expanded substantially. Nevertheless, many 'top-down' projects do attempt to create mechanisms for the effective participation of local communities. The Visakhapatnam Urban Community Development (UCD) Project in Andhra Pradesh, India, is based upon the principles of community organization, initiative, self-help and mutual aid. Community organizers establish personal contact with slum dwellers and encourage the development of neighbourhood committees, the members of which act as representatives for ten to fifteen families. These organizations (which tend to comprise only women) are responsible for co-ordinating a range of economic, educational and social activities at the community level. These include income-generation loans, the beneficiaries of which are selected by the neighbourhood organization; income-generation activities, such as co-operatives producing detergents and tooth powder; and sewing and adult education classes, which are funded by both community and project contributions. For the most part, community involvement in the wide range of promotional and preventive health activities is limited to providing venues and disseminating information. However, dwellers participate in the selection of community health volunteers, midwives and crèche asssistants. Finally, the UCD project aims to improve the environmental conditions of the city's slum dwellers by providing physical infrastructure and loans for self-help housing. Dwellers are consulted about the design and layout of physical improvements and are trained to maintain and carry out simple repairs of bore-wells, water pipelines, latrines and storm-water drains. In an attempt to raise women's security and status, negotiations about the housing scheme are increasingly channelled through the neighbourhood organizations, and land titles and loans issued in women's names.

The nature of the implementing agency

Most reviews of participation programmes make the distinction between the policies and outcomes of governmental, multilateral and non-governmental organizations (NGOs). Multilateral organizations such as the World Bank and UNICEF tend to give loans or grants to governmental agencies to implement large-scale projects. Having access to considerable financial resources, they may place political or bureaucratic pressures on the implementing agency to justify its policies and expenditure and to give tangible evidence of success. Constrained by both national and international considerations, government-funded projects are inevitably 'top-down' in orientation.

NGOs vary widely in terms of their size, resources and power. However, international agencies such as Oxfam and Save the Children prefer to give financial and technical assistance to local NGOs, many of which are religious or political in orientation. Non-government-funded projects are generally much smaller in scale and less bound by red tape and bureaucracy than official projects. This allows a greater freedom and flexibility to respond to local needs and initiatives and to experiment with new techniques.

The role and training of project staff

Deprofessionalization is a key theme in community participation literature, which states that medical doctors must cast aside the traditionally élitist attitudes of their profession and be open to learn from community people who have less education and expertise than themselves. A central role is given to the community health worker (CHW), a local resident who receives elementary training in preventive health measures, health education and simple curative care. Under *means* definitions of participation, CHWs have a *service* role as extenders of primary health services. More radical definitions give CHWs a *developmental* role as *change agents*. Here, they are envisaged as catalysts who help communities to understand about ill-health and the factors causing it and to identify solutions to their situation.

In practice, few CHW schemes have lived up to their expectations. Rather than being selected by and accountable to the communities that they serve, community leaders and health professionals usually identify CHWs, who come to be regarded as low-level government employees. CHWs rarely represent the poorest sections of communities, and the tendency of some national CHW programmes (for example, in India) to set minimum educational requirements can result in the exclusion of women. Other tensions exist between the amount of time given to curative tasks, preventive work and promoting participation. The former often take precedence, especially when CHWs are based in health centres, which can result in the CHW's work becoming almost completely confined to the health sector. Finally, CHW schemes have been undermined by a lack of financial commitment which has resulted in inadequate training, supervision and support (including drugs and supplies) and low staff morale (Walt 1990).

Mechanisms for promoting participation

Participatory projects are invariably channelled through some form of local organization. Although traditional associations such as tribal clans and caste-based organizations exist throughout the Third World, they are often condemned for representing élitist interests. In response, project planners have attempted to set up new organizational structures, a popular example of which is the village health committee (VHC). VHCs should comprise a representative selection of community members. Typically, their functions include co-ordinating with the implementation agency, gathering information about the local health situation, selecting priority problems and solutions, assigning responsibilities for different activities and mobilizing community resources.

In practice, the methods by which such committees are created and the structures and functions imposed upon them suggest that, rather than being expressions of felt needs and community-defined solutions, they are outcomes of bureaucratic demands. Consequently, participation efforts are quickly routinized and harnessed to agency goals and procedures. Health planners are today reassessing the value of creating new institutions and concluding that it is better to strengthen and work within existing leadership and community organization structures. In Korea, for example, indigenous women's groups have been formalized into mothers' clubs which undertake family-planning, income-generation and credit activities.

The question of whether to use a traditional organization or to create a modern one clearly depends upon the local context. Where communities are characterized by factionalism, patronage or concentrated leadership, development agencies are likely to demand the formal establishment of representative organizations.

Community characteristics

There is no guarantee that even the best-planned project can result in permanent and dynamic community involvement. Because many 'communities' in the geographical sense are economically, socially and politically heterogeneous, different interests may conflict and compete for project resources. The poorest community groups are usually identified as potential participants in social welfare programmes. However, pressures easily build up to steer benefits towards the less poor. There are a number of reasons for this. First, the poor are the least able to afford the time required to participate in community activities. Secondly, local leaders may represent wealthier groups and act in their interests. One major problem is the lack of female leadership in many Third World communities, which, it is argued, presents an obstacle to the filtering through of project benefits to women and to their participation itself.

Community leadership is perhaps the most important factor influencing the participation process at the local level. A democratic and representative leadership structure is a prerequisite of broad-based and effective participation, for although charismatic leaders may mobilize people and resources effectively, the emergence of strong and enduring collective institutions is retarded. Unfortunately, support systems based on collective and consensual decision-making are virtually unknown. The nature of the resource provision system in many less developed countries means that local leaders play a key role in bargaining with political and governmental patrons for community services. As their communities reward them with power, prestige and financial recompense, it is not in their interests to encourage mass participation.

CONCLUSION

This chapter began with a brief description of the acrimonious debate on comprehensive PHC and SPHC. Supporters of the former believe that community participation has an active role to play in bringing about the equitable distribution of power and resources. SPHC, on the other hand, implies a more limited definition of participation, whereby community involvement is sought to increase the coverage and/or effectiveness of specific health interventions. Those who argue that the ideas at the core of PHC are revolutionary suggest that SPHC represents a backward step in the evolution of international health policy. However, in the new political and economic climate, vertical selective programmes have taken precedence over longer-term development strategies.

Ironically, the emphasis in the primary health care literature on the participation of individual communities unintentionally contributed to this ideological shift. First, the romantic rhetoric of 'people power' detracted attention from the macro-scale causes of underdevelopment which lie in the world economy. Secondly, support for community self-reliance is quite compatible with the neo-classical conviction that state intervention stifles initiative, engenders dependency and obstructs capitalism. As a result, the concept of participation became a double-edged sword, providing a popular language for policies which are radically different to those envisaged at Alma Ata. With less than a decade to go, the goal of 'Health for all by the year 2000' remains a distant dream.

NOTES

1 The Alma Ata conference declared that PHC should include education about prevailing health problems and methods of preventing and controlling them; promotion of food supply and proper nutrition; an adequate supply of safe water and basic sanitation; maternal and child health, including family planning; immunization against infectious diseases; prevention and control of endemic diseases; appropriate treatment of common diseases and injuries; and provision of essential drugs.
2 Under what is referred to as the GOBI strategy, UNICEF identified Growth monitoring, Oral rehydration therapy, Breast-feeding and Immunizations as specific measures for improving child health. The strategy was later expanded to include Family spacing, Female education and Food supplements (GOBI–FFF) (see Chapter 9).

REFERENCES

Agudelo, C.A. (1983) 'Community participation in health activities: some concepts and appraisal criteria', *Bulletin of the Pan American Health Organisation* 17(4): 375–85.
Hollnsteiner, M.R. (1982) 'The participatory incentive in primary health care', *Assignment Children* 59/60: 35–56.
Morley, D., Rohde, J. and Williams, G. (eds) (1983) *Practising Health for All*, Oxford: Oxford University Press.
Oakley, P. (1989) *Community Involvement in Health Development: An Examination of the Critical Issues*, Geneva: WHO.
Rifkin, S. (1985) *Health Planning and Community Participation: Case Studies in South-East Asia*, London: Croom Helm.
—— (1990) *Community Participation in Maternal and Child Health/Family Planning Programmes*, Geneva: WHO.
Rifkin, S. and Walt, G. (eds) (1986) Special edition on Selective or Comprehensive Primary Health Care, *Social Science and Medicine* 26(9).
Shah, P.M. (1976) 'Community participation and nutrition: the Kasa project in India', *Assignment Children* 35: 53–71.
Unger, J.P. and Killingsworth, J.R. (1986) 'Selective primary health care: a critical review of methods and results', *Social Science and Medicine* 22(10): 1001–13.
Walsh, J.A. and Warren, K.S. (1979) 'Selective primary health care: an interim strategy for disease control in developing countries', *New England Journal of Medicine* 301(18): 967–74.
Walt, G. (ed.) (1990) *Community Health Workers in National Programmes*, Milton Keynes: Open University Press.

Part III

REGIONAL ISSUES IN HEALTH AND DEVELOPMENT

13

THE POORER THIRD WORLD

Health and health care in areas that have yet to
experience substantial development

Helmut Kloos

The poorer Third World, also known as 'low-income economies' and 'least
developed countries' (referred to here as LDCs), includes nearly all nations
in sub-Saharan Africa, many countries in South and South-east Asia, China,
India, and Haiti and Guyana in Latin America. But the world's least
developed areas, which are characterized by the highest incidence of
preventable communicable diseases and malnutrition, also include numerous
impoverished areas and communities within middle-income and high-
income countries, such as the Bantustans and black ghettos in South Africa,
many rapidly growing city slums and socially and economically marginalized populations, including the Australian Aborigines, South American
indigenous peoples, peoples in polar areas, many pastoral nomads, low
castes, refugees and other displaced groups (Swedlund and Armelagos 1991).
According to the World Bank, including China and India, 3.1 billion of the
world's 5.5 billion people in 1990 lived in the forty-three low-income
countries with gross domestic product (GDP) per capita of $610 or less.
These countries constitute a large variety of physical and biotic environments inhabited by similarly diverse racial, ethnic and socio-economic
groups and political institutions, all of which interact to produce a multitude
of health and disease states that are only gradually being comprehended
(Polednak 1989).

The LDCs share the lowest quality of life indices, particularly the highest
mortality and morbidity rates, extreme poverty and slow, if not stagnant,
socio-economic growth. Sub-Saharan Africa, the poorest region, even
experienced an annual decline of gross national product (GNP) per capita
of 1.2 per cent during the 1980s (World Bank 1991) and saw six of its countries
moved from middle-income to low-income status. In India and Bangladesh,
in spite of moderate growth in national GNP during the 1970s, the percentage
of people in poverty increased, reflecting inequitable development.

The objective of this chapter is to examine socio-economic, epidemiological and political aspects of disease distribution and health development

199

in the LDCs and other least developed areas. Considerable attention is given to the role of poverty and disasters in ill-health, two key factors often neglected in epidemiological and medico-geographical studies.

POVERTY, HEALTH AND SUSTAINABLE DEVELOPMENT

Poverty, a broad concept that encompasses many aspects of life and the social and physical environment, is perhaps the major determinant of ill-health in LDCs. There were more than 1.1 billion poor people, including 630 million absolutely poor people (annual per capita consumption below $275) in the LDCs in 1985, nearly half of them in South Asia, one-quarter in East Asia and one-fifth in Africa (World Bank 1991). The number of poor and the severity of poverty have increased during this century, largely due to the convergence of structural poverty (mainly the result of long-term correlates such as landlessness) and conjunctural poverty (due to drought and other short-term misfortunes). Famine prevention programmes have not solved the chronic food crisis in Africa and South Asia. As a result, epidemic starvation has gradually been replaced by massive endemic undernutrition of the poor. More holistic approaches to reducing poverty and solving the food crisis (Dreze and Sen 1990) may eventually lead to a reversal of the present downward trend. In the meantime, the well-known association between nutrition and immunity has many implications for infectious disease occurrence.

Poverty manifests itself in many forms impacting on health, including poor housing, environmental sanitation and water supply, uncontrolled vector occurrence, unemployment and underemployment, low education achievement, high morbidity and mortality and poor access to health services. Other variables that impact, often indirectly, on health and socio-economic well-being include rapid population growth, low production, lack of political will to allocate necessary resources for social services and lack of infrastructure. Modelling of the relationship between national health status and health determinants has shown that, in countries with the lowest health levels, poor economic and governmental infrastructure, rather than health services, is most closely associated with health. In the absence of adequate health services, economic factors, education and government infrastrucure become the major determinants of health status (Hunter 1990).

The poor economic performance of the LDCs in the 1980s was associated with declining food production (0.8 per cent per annum) and industry and manufacturing (0.2 per cent); with declines in per capita food consumption and nutritional status, in spite of increased food imports; and with deterioration of health services and school enrolment (United Nations 1989). Some of the root causes of this calamity are extremely high population growth rates (2.7 per cent annually, excluding China and India [Table 13.1]), protracted wars, policy failures, mismanagement, inequitable entitlement

Table 13.1 Economic, demographic and nutrition indicators in the forty-one LDCs

Country	GNP per capita (US$)[a]	GDP annual average growth rate (%)[a] (1980–9)	Daily per capita calories supply as percentage of requirements (1984–6)[b]	Average annual growth of population (1980–90)[a]
All low-income countries	330	6.2	no data	2.0
All low-income countries without China and India	300	3.4	no data	2.7
Mozambique	80	–1.4	69	2.7
Ethiopia	120	1.9	71	3.0
Tanzania	130	2.6	96	3.1
Somalia	170	3.0	90	3.0
Bangladesh	180	3.5	83	2.6
Bhutan	180[c]	8.1	no data	2.1
Laos	180	no data	104	2.7
Malawi	180	2.7	102	3.4
Nepal	180	4.6	93	2.6
Chad	190	6.5	69	2.4
Myanmar (Burma)	220[c]	no data	119	2.1
Burundi	220	4.3	97	2.9
Sierra Leone	220	0.6	81	2.4
Madagascar	230	0.8	106	2.9
Vietnam	240[c]	no data	105	2.1
Nigeria	250	–0.4	90	3.4
Uganda	250	2.5	95	3.2
Zaire	260	1.9	98	3.1
Mali	270	3.8	86	2.5
Afghanistan	280[c]	no data	94	–0.3[d]
Niger	290	–1.6	100	3.4
Burkina Faso	320	5.0	86	2.6
Rwanda	320	1.5	81	3.2
India	340	5.3	100	2.1
China	350	9.7	111	1.4
Haiti	360	2.9	84	1.9
Kenya	360	4.1	92	3.9
Pakistan	370	6.4	97	3.2
Benin	380	1.8	95	3.2
Central African Republic	390	1.4	86	2.7
Ghana	390	2.8	76	3.4
Togo	390	1.4	97	3.5
Zambia	390	0.8	92	3.7
Guinea	430	no data	77	2.5
Sri Lanka	430	4.0	110	1.5
Liberia	450[c]	no data	102	3.2
Lesotho	470	3.7	101	2.7
Sudan	480[c]	no data	88	2.8
Indonesia	500	7.0	116	2.1
Mauritania	500	1.4	92	2.4
Cambodia	no data	no data	98	2.6[d]

Sources: [a] based on World Bank (1991); [b–d] based on UNICEF (1991)

and poor distribution systems, and intensification of land use beyond the carrying capacity of the land (Glantz 1987; Dreze and Sen 1990). External factors, above all the world economic recession and the debt crisis of developing countries, contributed significantly to the deterioration of living and health conditions, through their negative effects on production, trade balances and government health budgets, free primary education, and food and fuel subsidies – the services the poor can least afford. Many of these conditions interacted with periodic droughts to accelerate ongoing environmental deterioration and ecological upheavals, such as those in the Sahel and the earlier ones in China and India (Glantz 1987).

Structural adjustment programmes were developed by the World Bank and the International Monetary Fund during the 1980s to deal with deteriorating living conditions in LDCs, by reducing their crippling international debts and stabilizing their economies, with anticipated consequent improvements in social conditions (Chapter 3). While several countries benefited from such growth-oriented programmes, these Draconian measures have also tended to jeopardize the wider social objectives of most governments, particularly poverty reduction, through budget cuts in health, education and nutritional support programmes. These stifling effects prompted in the mid-1980s the call for greater protection of the poor and other vulnerable segments of society through focused health, education and nutrition programmes as a prerequisite for sustained economic and social development. The potential impacts of structural adjustment programmes on health, particularly for the most vulnerable and those in the poorest countries, have been touched on in Chapters 1 and 3. The moral appeal of these and other basic-needs approaches is also evident in the declaration of 'Health for all by the year 2000', Sen's entitlement approach to food acquisition by the poor (Dreze and Sen 1990), child survival strategies led by the United Nations Children's Fund (UNICEF's) GOBI programmes and the World Declaration on the Survival, Protection and Development of Children (UNICEF 1991). More attention is now also being given to the environment; and ecologically sustainable development, rather than mere growth, is increasingly becoming the official policy of national and international organizations, including the World Bank.

PATTERNS AND TRENDS OF MORTALITY AND MORBIDITY

Health indicators

Indicators of health in LDCs are difficult to develop and have been of limited use due to the scarcity of relevant data and difficulties of interpreting morbidity and mortality in hyper-endemic multiple-disease environments. Although GNP per capita is in a general way related to health status, it is

Table 13.2 Under-5 and infant mortality rates (1969 and 1989), GNP per capita
and life expectancy at birth (1989), LDCs

Country	Under-5 mortality rate		Infant mortality rate		GNP per capita (US$)	Life expectancy (years)
	1989	Per cent decrease from 1960	1989	Per cent decrease from 1960		
Mozambique	297	10	173	17	80	47
Afghanistan	296	22	169	21	280	42
Angola[a]	292	15	173	17	610	45
Mali	287	22	166	21	270	45
Sierra Leone	261	32	151	31	220	42
Malawi	258	30	147	29	180	48
Guinea	241	28	142	30	430	43
Burkina Faso	232	36	135	34	320	48
Ethiopia	226	23	133	24	120	45
Niger	225	30	132	31	290	45
Central African Republic	219	29	129	30	390	49
Chad	219	33	129	34	190	46
Somalia	218	26	129	26	170	46
Mauritania	217	32	124	35	500	47
Liberia	209	33	137	26	450	53
Rwanda	201	19	119	18	320	49
Cambodia	200	8	127	13	no data	50
Burundi	196	25	116	24	220	48
Bhutan	193	35	125	33	180	49
Nepal	193	35	125	33	180	52
Yemen[a]	192	48	116	46	650	51
Senegal[a]	189	48	85	51	650	48
Bangladesh	184	30	116	26	180	51
Madagascar	179	51	117	47	230	54
Sudan	175	40	105	38	480	50
Tanzania	173	31	103	30	130	54
Namibia[a]	171	35	103	34	1030	57
Nigeria	170	42	102	46	250	51
Gabon[b]	167	42	100	42	2960	52
Uganda	167	25	100	23	250	52
Bolivia[a]	165	41	105	37	620	54
Pakistan	162	41	106	35	370	57
Laos	156	33	106	32	180	49
Benin	150	52	89	52	380	47
Cameroon[a]	150	45	92	44	1000	53
Togo	150	51	92	49	390	54
India	145	49	96	42	340	59
Ghana	143	36	87	34	390	55
Total very high U5MR countries	193	35	118	36	no data	49
Total high U5MR countries	94	59	67	53	no data	61
Total middle U5MR countries	36	70	27	69	no data	70

Sources: Based on UNICEF (1991); World Bank (1991)
[a] Middle-income country.
[b] Upper-middle-income country.

not a sensitive indicator of health. Relatively high levels of GNP per capita do not necessarily result in better health of the population, and vice versa; some countries with low GNP indices have achieved remarkably high health levels (Table 13.2), further discussed below. Infant mortality has been widely used as a fairly reliable measure of child health and development by the international donor community, for the purpose of monitoring progress in child survival programmes. However, while reflecting quite well maternal and neo-natal conditions, it is a less precise indicator of the environment necessary for satisfactory growth of older children, let alone adults. UNICEF and other donor organizations are increasingly using the under-5 mortality rate (U5MR) as a more sensitive health indicator.

All mortality measures in LDCs are deficient, however, in that they fail to link mortality rates satisfactorily to cause of death, and because of under-reporting. In Africa, only 0.25 per cent of all deaths are being reported to the World Health Organization (WHO), compared to 4 per cent in South-east Asia and 94 per cent in Europe. Morbidity statistics are similarly incomplete. Differential access to health services among different social, cultural and residential groups and differences in medical technology, diagnostic techniques and reporting render existing morbidity and mortality statistics non-comparable. Moreover, many national averages are not representative in view of the great differentials between socio-economic groups and geographical areas. UNICEF, the WHO and other organizations by necessity are utilizing morbidity and mortality estimates based on demographic structure, health services provided (e.g. immunization coverage) and community health surveys. A new index of preventable infant and child deaths which links mortality levels with cause-of-death structures and considers the progressively increasing difficulty of reducing mortality levels has been recommended (D'Souza 1989). Life expectancy at birth, based on demographic parameters, is widely used in LDCs as a convenient aggregate health indicator for all age groups.

A note on morbidity

Although comparative analysis of health status of LDCs is problematic, available data on major diseases do permit the search for causative factors and interventions in the quest for health improvement at affordable cost and with available technology and other domestic resources. The prevalence and geographical distribution of major infectious and nutritional deficiency diseases in the developing world have been mapped by many authorities. These studies reveal a preponderance of the major killing and debilitating communicable diseases and nutritional disorders in sub-Saharan Africa and to a somewhat lesser degree in the other least developed areas. AIDS, earlier thought to be rare in much of Asia, assumed epidemic proportions in South and South-east Asia around 1990. The rapid spread of drug-resistant malaria and diarrhoea, as well as rapid urbanization, industrialization and water resources development, are often exacerbating rather than solving health problems.

Infant and under-5 mortality

Infant mortality constitutes around 60 per cent of all deaths in LDCs, compared to 2 per cent in developed countries. Infant mortality and under-5 mortality rates (U5MRs) for the thirty-eight countries with the highest U5MRs in 1989 are shown in Table 13.2. U5MRs were twice as high in countries with the highest rates (Mozambique, Afghanistan and Angola, with 297, 296 and 292 deaths per 1,000 live births) as in countries with the lowest rates, Ghana and India (145 and 143 per 1,000). Thirteen of the fifteen countries with the highest rates were in sub-Saharan Africa. However, nearly half of all children dying below age 5 were in India, due to its large population. Infant mortality followed a similar pattern. At the regional level, the highest U5MRs were in Africa (166 per 1,000), particularly in the very-high-U5MR countries (203), followed by South Asia (155), South-east Asia (92), South-west Asia (71), Latin America (68) and the developed countries (11).

These high mortality rates are due primarily to preventable communicable diseases and malnutrition, reflecting the poverty-stricken environment of LDCs. Most children in LDCs experience more than one episode of life-threatening disease, often simultaneously, making the classification of cause of death and the selection of intervention strategies difficult. Because high exposure levels or disease attacks continue into adulthood, much of the relatively quick success achieved by short-term control programmes in reducing infant and child mortality is negated by subsequent disease attacks. This dilemma re-emphasizes the need for combined, well-coordinated health and socio-economic development of LDCs.

The top seven causes of death in children under 5 reported by developing countries in 1985 were acute respiratory diseases (causing an estimated 5 million deaths), diarrhoeal diseases (4 million deaths), perinatal causes, including low birth-weight (3.2 million), malaria (750,000), tuberculosis (300,000), other infectious and parasitic diseases (450,000) and other and unknown causes (700,000). Nutritional disorders in infants and children under 5, including moderate and severe underweight in infants (17–71 per cent prevalence in the thirty-eight very-high-U5MR countries), wasting in 1–2 year-olds (2–18 per cent) and stunting in 2–4 year-olds (25–70 per cent), must be expected to have greatly contributed to these high mortalities. The inadequacy of the staple foods, as reflected in the low calorie supply (Table 13.1), is much to blame for this situation, but avitaminoses, above all vitamin A deficiency, also contributed to mortality (UNICEF 1991).

While UNICEF came close to achieving its goal of 80 per cent immunization coverage in infants in developing countries by 1990, only 47 per cent of infants in LDCs had been vaccinated for DPT (diphtheria, pertussis (whooping cough), tetanus) by 1989 and 38 per cent against measles. Coverage was only 18–46 per cent in Ethiopia, Angola and Afghanistan, the lowest achievers. Similarly, only 32 per cent of pregnant women had been immunized against tetanus and 23 per cent used oral rehydration therapy (ORT). Overall immunization coverage in the thirty-eight countries with the highest U5MRs was only two-thirds of that in countries with high U5MRs and slightly over half of that in middle-income countries. Access to safe water and health services showed a similar pattern, but with only 45 per cent coverage in these thirty-eight countries. Increases in coverage in the 1980s were much greater for immunization (25–183 per cent) for the different vaccines than for water supply (4 per cent in urban areas and 10 per cent in rural areas) (UNICEF 1991). These differences are largely due to the relatively greater success of targeted, often vertical, immunization programmes than water supply and sanitation programmes, which depend more on broader socio-economic development. The poor performance of the water and sanitation campaign of the 1980s is all the more unfortunate since their implementation can boost immunization coverage through complementary health programmes and infrastructural measures.

The data on infant mortality and U5MRs in Table 13.2 reveal two patterns with causal implications. First, the largest declines in both types of mortality between 1960 and 1989 were achieved by the lowest-U5MR countries. This disturbing finding is corroborated by the even greater reduction of U5MRs (by 136 deaths per 1,000 children, on average) and infant mortality (76 per 1,000 live births) by the high-U5MR countries (UNICEF 1991) during the 1960–89 period, again revealing the failure of the least developed countries to make substantial progress in health status. U5MRs may even increase in many LDCs, especially among millions of orphans whose parents are expected to die of AIDS. Secondly a number of low-income countries (Zaire, Zambia, Indonesia, Myanmar, Vietnam, Sri Lanka and China) in Table 13.1 were not even among the top forty-one U5MR countries; and, vice versa, seven middle-income countries were in the highest U5MR group. This re-emphasizes the weakness of GNP per capita as a health indicator.

The rapid decline in infant and child mortality that led to high life expectancy relative to GNP per capita in Sri Lanka, Costa Rica, China and Kerala State in India (Tables 13.1 and 13.2) has been widely discussed in the literature as a model of health development in LCDs. These declines have been associated with the egalitarian political system of these countries, a high social status for women, a strong demand for education and health services, a mechanism to guarantee the efficient operation of health services, an adequate minimum standard of nutrition, universal immunization and strong state initiative.

Adult mortality

The still sketchy data on adult mortality reveal a distribution pattern somewhat different from that of infant and child mortality. Whereas the highest rates persist in sub-Saharan Africa and rates have declined (together with child mortality) in China, Sri Lanka and Kerala State during the last three decades, there is little difference in adult mortality rates between some African LDCs and a number of large Latin American countries in the middle-income category. Chronic diseases often associated with ageing – as well as tuberculosis, one of the most neglected and most difficult-to-treat diseases in LDCs – dominate the cause-of-death structure of adults, with AIDS rapidly becoming a major killer in many LDCs. Tuberculosis is a major opportunistic infection in HIV-positive people, posing a major challenge for LDCs. Circulatory and other degenerative diseases as well as injuries are becoming major causes of morbidity, disability and mortality in LCDs (Feachem et al. 1992).

Significant changes have occurred in traditional societies in LDCs in recent years, including rapid urbanization. Industrialization and the adoption of modern lifestyles, such as cigarette smoking, alcohol consumption and refined diets, are associated with an increase in chronic non-infectious diseases. An estimated 1.8 million out of the 2.3 million deaths due to chronic obstructive respiratory diseases in developing countries in 1985 occurred in India and China (only 60,000 in sub-Saharan Africa), as well as 1.4 million of the 2.5 million cancer deaths. The 1985 incidence of 6.5 million deaths from circulatory diseases in developing countries was estimated by the WHO primarily on the basis of age structure of the population in the absence of epidemiological data.

Communicable diseases and maternal mortality continue to dominate the mortality pattern in the 15–44 age group in LDCs. Maternal mortality and morbidity, still poorly understood, appear to be much higher than official statistics, typically gathered by hospitals, indicate. In several populations in Africa and in Indonesia complications of pregnancy and childbirth are the major or second major cause of death in 15- to 44-year olds. Family planning can play an important role in reducing maternal mortality.

Disaster

Health aspects of natural and human induced disasters in developing countries have been relatively neglected, largely because of scarcity of reliable data and the interdisciplinary nature of such studies. Drought, floods, earthquakes, storms, cyclones, famines and epidemics have significantly greater health impacts in the LDCs than in industrialized countries, primarily because of the inability of the former to prevent disasters or ameliorate their effects. This relationship is illustrated by the history of drought, famine and floods in China, where devastating disasters could be prevented only after the revolution in 1949, when the country achieved greater economic security.

Between 1961 and 1981, natural disasters (cyclones, earthquakes, drought and floods) claimed 1.1 million lives in low-income countries, 345,000 in middle-income countries and only 11,300 in high-income countries, although the number of disasters experienced by these three groups of countries was similar (Desai 1990). For the period 1980–90, the WHO's Pan-African Centre for Emergency Preparedness and Response reported 172,000 deaths from epidemics (mostly due to malaria, yellow fever, meningitis and cholera), 6,613 deaths due to floods, cyclones and earthquakes, 10.6 million people affected by these various disasters, 566,000 deaths (a gross underestimate) from famines, drought and food shortages, and 9.3 million homeless people and refugees. All these figures are underestimates due to underreporting. Often different disasters interact and, when complicated by war and civil disturbance, result in extremely high mortality. In four of the six African countries experiencing extreme food shortages in the mid-1980s (Ethiopia, Chad, Mozambique and Sudan), drought and externally financed insurgency triggered severe famine, after locally high population growth, economic collapse and ecologically damaging land use had brought the land to the verge of environmental bankruptcy (Glantz 1987). In different countries, particularly in the highlands of northern Ethiopia, this situation has reduced the carrying capacity of the land and was instrumental in the occurrence of more frequent and severe famines and out-migration. In developing countries worldwide there are an estimated 30 million refugees and people displaced by war, civil unrest and persecution. Mortality rates in these populations are up to sixty times greater than the expected rates, peaking in children aged under 15 years (Toole and Waldman 1990).

The effects of war in LDCs are similarly underreported. Most countries with the highest U5MRs in 1989 and the lowest declines in rates – including Mozambique, Afghanistan, Angola, Mauritania, Guinea, Chad, Sudan, Ethiopia, Rwanda, Burundi, Somalia, Uganda, Laos and Cambodia – have been engaging in protracted wars or bloody revolutionary struggles. In addition to contributing to high mortality and morbidity, these wars disrupted food production systems and trade, reduced tax revenues, destroyed public health services and schools and diverted scarce money into the military (up to 50 per cent of national budgets) (Sivard 1991). Many military governments have also been charged with state violence against the public (Alubo 1990; Sivard 1991).

Another form of disaster, state-based racial discrimination, has caused excessive morbidity and mortality in South Africa, a middle-income, middle-health-level country, due to demographic and epidemiological polarization. The health and socio-economic status of its black majority population are similar to those of LDCs. Infant mortality rates in blacks in 1981 were as high as 130 (in Transkei) and 144 (in Kwazulu) but were only 12 in whites. Rates have not decreased throughout the country, having increased in several poor areas, including Siskei and Transvaal (Wilson and Ramphele 1989).

HEALTH CARE

The tremendous health needs in the LDCs, as indicated by their low health levels and growing demands by the public, obviously call for the expansion and strengthening of preventive and curative health services. The official health services in many countries are facing difficulties in meeting basic health needs, largely because of lack of resources. The 'brain drain' of health personnel is contributing to this problem in some areas. But the low quality of many health services and the availability of alternative care, together with strong social, cultural and physical barriers between patients and the medical services, have been instrumental in their underutilization in many LDCs. The economic crisis has contributed to the health care problems.

Medical pluralism

Pluralistic medical systems, found in practically all LDCs, consist of a plethora of traditional healing systems, lay practices, household remedies, transitional health workers (usually lay persons practising modern medicine illegally) and Western biomedicine. Although most traditional medical systems and practices continue to be widely used and constitute the only health resource in many isolated areas, they have not been fully recognized by or integrated into the official health systems of most LDCs, particularly in Africa (see Chapter 4). In China, by contrast, serious attempts have been made to integrate the two systems. Traditional birth attendants are the only component of traditional medicine that has been fairly widely utilized in national health systems, as part of the primary health care (PHC) initiative.

Introduced during the colonial era, modern health services in LDCs were initially operated by missions and colonial administrations. After independence, some newly established national health systems continued to adhere to a large extent to colonial administrative practices, while more radical governments adopted socialist approaches. After an early rapid expansion of urban-based curative systems, which contributed to raising health levels mostly in urban areas, nearly all LDCs adopted health care policies based on the PHC approach, with variable success.

Problems faced by the modern health services

Major problems that Western biomedicine has faced in LDCs include: (1) a failure to direct health policies towards the major causes of morbidity and mortality in the general population, instead of towards the health problems of the privileged, urban-based élite; (2) difficulties of planning agencies in implementing PHC programmes; and (3) a lack of resources, including community participation and intersectoral collaboration. Health expenditures constitute only about 3 to 4 per cent of government budgets in most low-income countries (but 5 to 6 per cent in middle-income countries). Planning, management and support structures of PHC have remained weak; intersectoral collaboration is difficult to achieve; and PHC is widely perceived by the medical community as an archaic, neo-colonial approach. It is even suggested by some that top-down planning in health services development in Africa, deeply rooted in African political economy, is jeopardizing the very survival of PHC. Other researchers have been more optimistic, considering PHC programmes to be potentially effective and merely in need of stronger support.

On the more positive side, while the economic depression of the 1980s was another impediment to the development of effective health services, it also provided the impetus for upgrading health programmes. Thus new and renewed emphasis has been placed on the reduction of disparities between urban and rural sectors and, within urban areas, particularly better coverage of vulnerable and underprivileged groups, and increased community participation. But the rapidly growing burden of caring for AIDS patients and the increasing health needs of the growing elderly population are causing increasing strains in medical services and will require further adjustments in health policies and planning, and in the provision of care.

Comparative study of health systems

Roemer (1991) developed the first typology of health systems in low-income countries, based on political ideology. He classified the health systems of selected LDCs into (1) entrepreneurial, (2) welfare-oriented, (3) comprehensive and (4) socialist systems. Salient features of the fourteen health systems selected by Roemer and of additional systems selected by the present writer are summarized in Table 13.3. They are considered to be fairly representative of LDCs.

Variations in the specific features of these health systems and the progress they have made can to a large extent be explained by differential growth of the economies and, perhaps even more so, by differences in the redistribution of wealth and political will. A common consensus in the redistribution debate is that, while economic growth is necessary for improvement of living standards and social services, countries emphasizing redistribution as a major development objective (and not only socialist countries) are characterized by better education, lower infant mortality and higher life expectancy than countries choosing growth alone. Relevant to the medico-geographical perspective is the association between redistribution levels and the distribution

Table 13.3 Summary data on health services in nineteen LDCs, classified by their political economy

Country	Population per physician (1987)[a]	Births attended by health staff[b] (%) (1987)	Percentage of pop. with access to health services[b] Rural	Urban	Contraceptive prevalence (1980-8) (%)[c]	Percentage of children vaccinated (1988-9)[c,d]	Percentage of pop. with access to safe water supply[b] 1985-8	Public health expenditure for health per capita (1987, US$)[c]
Socialist:								
Vietnam	1,309	99	80	100	20	68-80	46	no data
China	1,670	no data	no data		74	95-98	85[f]	4
Mozambique	43,803	28	30	100	47	32-51	16	2
Ethiopia	73,083	58[g]	46[h]		2	23-44	16	2
Afghanistan	5,884	no data	17	80	2	22-38	21	no data
Angola[e]	18,452	15	30[h]		1	18-46	30	11
Comprehensive:								
Sri Lanka	7,202	87	93[h]		62	81-97	40	7
Tanzania	20,407	74	72	99	1	82-93	56	1
Welfare-oriented:								
Myanmar (Burma)	3,534	93	11	100	5	45-66	49	2
India	2,621	39	30[l]	75[l]	34	69-89	57	3
Zimbabwe[e]	6,801	69	62	100	43	70-80	32[i]	20
Liberia	9,671	89	30	50	6	28-62	55	8
Kerala State	2,389	72	91[l]	100[l]	no data	90-95[j]	57	4
Entrepreneurial:								
Kenya	9,627	no data	30[k]	92	27	62-90	30	7
Ghana	7,213	73	45	40	13	51-99	56	4
Zaire	13,079	no data	17	99	1	38-54	33	1
Pakistan	2,266	24	35		8	64-78	44	1
Indonesia	8,009	43	80[h]		48	68-85	38	2
Bangladesh	6,587	no data	45[h]		25	49-89	46	1

Sources and notes:
a Based on Sivard (1991).
b,c Based on UNICEF (1991).
d Numbers show the range of percentage coverage for the six immunizable childhood diseases.
e Middle-income country.
f Urban areas only.

g Gross overestimate: different studies indicate 20-25% coverage.
h Rural and urban.
i Rural areas only.
j Estimated by the author, based on various reports.
k Urban and rural, based on Roemer (1991).
l 1979 data, based on Franke and Chasin (1989).

of health services and accessibility of care. None of the LDCs discussed here achieved both substantial growth of GNP per capita and redistribution. Even among middle-income countries, this goal is seldom achieved; probably only Korea and Taiwan and small city-states such as Hong Kong and Singapore having been able to achieve it.

Overall, there is considerable variation among health system categories and individual countries with regard to health inputs (such as public expenditures for health per capita, population per doctor and percentage of the population with access to care and safe water) and outputs (life expectancy, morbidity and mortality). Countries with life expectancy levels approaching those in developed countries have welfare-oriented (Kerala), comprehensive (Sri Lanka) or socialist (China) systems.

Socialist systems

Although a number of features are shared by the six socialist health care systems in Table 13.3, including the overriding role of the state in the planning and provision of health care, those in China and Vietnam are unusual in regard to the relatively large number of auxiliary health personnel, emphasis on PHC, highly structured communal health services and fully integrated traditional medicine. The higher life expectancy in China than in Vietnam is apparently due to its longer history as a socialist nation and to the toll taken by the wars in Vietnam between 1945 and 1975. Nevertheless, only about 40 per cent of the Chinese people had medical insurance in the later 1980s, curtailing the access to modern care of most rural people (Roemer 1991). Mozambique, Angola, Ethiopia and Afghanistan, all with a history of protracted revolutionary and counter-revolutionary wars, famine and severe poverty, have made only moderate achievements in developing their health services and raising health levels. The extremely low birth attendance rate, health services coverage, immunization coverage and life expectancy in Angola – a middle-income country that allocated a large proportion of its budget (11 per cent) for health – illustrates the disruptive effect of war.

Comprehensive systems

The higher health input and output levels of the Sri Lankan than the Tanzanian health services has been attributed to a combination of the longer period of independence, the egalitarian policies on health services, education and food distribution of the Sri Lankan government and the slow growth of the Tanzanian economy. Tanzania's commendable achievements in health care coverage, birth attendance, immunization and provision of safe water supplies can be traced to the Arusha Declaration issued in 1967, which called for egalitarianism and self-reliance. Although both countries initially

THE POORER THIRD WORLD

followed a socialist path of development and still provide free care for all, the system in Sri Lanka provides private health services (particularly Ayurvedic medicine). Ayurvedic medicine includes elements of Western medicine such as antibiotics and may thus contribute significantly to the remarkable high life expectancy in Sri Lanka (Roemer 1991).

Welfare-oriented systems

Of the countries with welfare-oriented systems, characterized by a pre-dominantly private sector (except Zimbabwe), with socialist influences and an emphasis on PHC, India and Kerala State are of particular interest. The significantly higher inputs and outputs of Kerala's health system are largely the result of good physical access to care and, more importantly, the egalitarian social services and land tenure system. In the rest of India, a combination of lower equity and difficulties with the PHC programmes has impeded progress towards stated health care objectives (Roemer 1991). The health system of Zimbabwe, a middle-income country, fared no better than those of Myanmar and Kerala, re-emphasizing the role of the redistribution factor. This is also evident in the Liberian system, which depends to a large degree on the provision of health services by foreign firms exclusively for their employees and families, and on costly services provided by missions (Roemer 1991).

Entrepreneurial systems

The health systems of Kenya, Ghana, Zaire, Pakistan, Indonesia and Bangladesh are the result of strong entrepreneurial policies. Although the Ministries of Health of these countries have developed nationwide networks for preventive and curative services, their support structures are weak, and health services are underutilized. Ambitious health plans have not been implemented, and the proportion of the Ministry of Health's share in the national government budget has been declining. The poor and other disadvantaged people are most affected. In Bangladesh, where coverage increased from 10 per cent in the mid-1970s to 45 per cent in 1988, a large proportion of 'modern' rural doctors are unqualified and unlicensed allopaths who, together with Ayurvedic practitioners, homeopaths, tradi-tional midwives, spiritual healers, bone-setters and other health workers, meet most health needs of the population (Akin et al. 1985). In Kenya and Pakistan most physicians, and in Ghana, Zaire and Indonesia most physicians with government appointments, spend much time in private practice. Most hospitals and medical schools in Indonesia are private. The relatively low coverage of water supplies, health services (especially in rural areas) and life expectancy in these countries point to the inadequacy of these health systems.

CONCLUSIONS

One of the inescapable conclusions of this chapter is that research and information systems need to be developed in LDCs with the aim of solving health problems in the long term. In addition to inequitable distribution of research funds between and within countries, most assistance (79 per cent) with all forms of development is bilateral, directly from one country to another; and only 5 per cent of this is spent on health problems of developing countries, where 93 per cent of the years of potential life are lost (CHRD 1990). The Commission on Health Research and Development (CHRD) recommended (similarly to the Brandt Report) that this situation be rectified by developing a programme of national health research for each country, with increased national and international support. The CHRD emphasized that health research is an important long-term investment in development that appreciates over time, whereas many short-term aid programmes depreciate. Simultaneous development of national health information systems is urgently needed for the purpose of evaluating and improving health services and monitoring progress towards health for all.

REFERENCES

Akin, J.S., Griffin, C.C., Guilkey, D.K. and Popkin, B.M. (1985) *The Demand for Primary Health Services in the Third World*, Totowa, NJ: Rowman & Allanheld.
Alubo, S.O. (1990) 'State violence and health in Nigeria', *Social Science and Medicine* 31: 1075–84.
Commission on Health Research and Development (1990) *Health Research: Essential Link to Equity in Development*, New York: Oxford University Press.
Desai, B. (1990) 'Managing ecological upheavals: a Third World perspective', *Social Science and Medicine* 30: 1065–72.
Dreze, J. and Sen, A. (eds) (1990) *The Political Economy of Hunger*, Oxford: Clarendon Press.
D'Souza, S. (1989) 'The assessment of preventable infant and child deaths in developing countries: some implications of a new index', *World Health Statistics Quarterly* 42: 16–25.
Feachem, R.G.A., Kjellstrom, T., Murray, C.T.L., Over, M. and Phillips, M.A. (eds) (1992) *The Health of Adults in the Developing World*, Oxford: Oxford University Press.
Franke, R. and Chasin, B.H. (1989) *Kerala: Radical Reform as Development in an Indian State*, San Francisco: Institute for Food and Development Policy.
Glantz, M.H. (ed.) (1987) *Drought and Famine in Africa: Denying Famine a Future*, New York: Cambridge University Press.
Hunter, S.S. (1990) 'Levels of health development: a new tool for comparative research and policy formulation', *Social Science and Medicine* 31: 433–44.
Polednak, A.P. (1989) *Racial and Ethnic Differences in Disease*, New York: Oxford University Press.
Roemer, M.I. (1991) *National Health Systems of the World, Vol. 1: The Countries*, Oxford: Oxford University Press.
Sivard, R. (1991) *World Military and Social Expenditures*, 10th edn, Washington, DC: World Priorities.

Swedlund, A.C. and Armelagos, C.J. (1991) *Disease in Populations in Transition*, Westport, CT: Greenwood Press.

Toole, M. and Waldman, R.J. (1990) 'Prevention of excess mortality in refugees and displaced population in developing countries', *Journal of the American Medical Association* 263: 3296–302.

United Nations (1989) *The Least Developed Countries: 1989 Report*, New York: UN.

United Nations Children's Fund (1987) *Children on the Frontline*, New York: UNICEF.

—— (1991) *The State of the World's Children 1991*, Oxford: Oxford University Press.

Wilson, F. and Ramphele, M. (1989) *Uprooting Poverty: The South African Challenge*, New York: Norton.

World Bank (1991) *World Development Report 1991*, New York: Oxford University Press.

14

SPATIAL INEQUALITIES AND HISTORICAL EVOLUTION IN HEALTH PROVISION

Indian and Zambian examples

Rais Akhtar and Nilofar Izhar

INTRODUCTION

There are wide disparities in the availability of welfare facilities including those for health at the international, national and regional levels. In terms of spatial distribution, such disparities often increase over time in both developed and developing countries. It has been said that health care provision is simply a manifestation of society's organization and distribution of scarce resources in space. In the Third World in particular, this distribution frequently leads to inequalities and lack of social justice, and produces effects which appear out of context and are often treated as individualistic health problems rather than measures of societal dysfunction. The problems encountered in providing health care in developing countries include shortage of trained personnel, inadequate preventive and curative care for large populations, shortages of drugs and exorbitant prices. These are in addition to the physical, socio-economic and cultural constraints encountered in the utilization of health care resources. The maldistribution of health care resources is one of the most serious problems facing all developing countries, and appears to affect especially those with a colonial past.

How widespread are such disparities globally and in which parts of the Third World specifically? In the Third World, the great majority of people suffer from excess mortality and morbidity, while, by contrast, the affluent enjoy a health status similar to that of most people in the developed world. Health services also show imbalances, with expenditure concentrated on sophisticated facilities in urban areas, leaving the rural majority practically unserved. From the health care perspective, the noble World Health Organization (WHO) slogan 'Health for all by the year 2000' has proven a great incentive to Third World countries, many of which are seriously concentrating on achieving this commitment, if with limited chance of success (Akhtar 1991).

216

Geographical research into health care generally focuses on aspects concerning facility and practitioner location, inequalities in accessibility, distance decay in utilization, travel time and costs. These aspects are analysed within the framework of the physical, socio-economic, cultural and political conditions of a given situation.

Health problems in developing countries such as India and Zambia are reflections of poverty, environmental contamination and inequity in the provision of health care in different regions. A substantial decrease in morbidity and mortality is likely to be accomplished by addressing these factors, and through an improved system for the distribution of health services. Therefore, the necessity of sufficient and equal distribution of health services such as health centres, hospitals, beds and paramedical and medical personnel can hardly be overemphasized.

THE HISTORICAL EVOLUTION OF HEALTH CARE IN INDIA

The present health facility structure in both India and Zambia is largely inherited from the colonial past. The colonial people practised selective development in order to serve their purposes. Port towns were developed, and the status of Western-type health care was raised even beyond the need of the select population, while the hinterlands were neglected. In the case of Zambia, the mineral-rich zone of the Copperbelt was the major focus of development. Later, selected urban, industrial and hill-resort centres were developed, with little consideration of needs-based health planning.

India has passed through various stages of health care development. During the colonial stage of development, the highest priority was accorded to the protection of the European population and to ensure that disease and epidemics did not affect Britain's trade. Only rarely did European physicians offer their services to local rulers and populations. Simultaneously, the indigenous or traditional systems of medicine that used to serve the vast majority of the population were greatly discouraged, and the Western type of medicine was introduced – again, in selected places. This has led to enclaves of development in terms of health facilities, and it may be said that health facilities are abundant where there is less need and vice versa.

The impact of colonial policy on health in India was contradictory. Changes in famine policy and food distribution helped to reduce mortality; increasing numbers of men (and, later, women) were trained in medicine to the international standards of the time; hospitals and dispensaries attracted considerable numbers of patients; and issues of disease prevention and public health provision were addressed as never before. However, many of these measures were restricted in their impact to a relatively small sector of the population: first, the European civil and military personnel and their families; and later, those with access to urban facilities. Preventive campaigns were limited, and the mass of the Indian population did not benefit (Jeffrey 1991). However, on a wider scale, Klein (1972) points out that the nature of the colonial economy and the ecological changes brought about (or hastened) under colonialism often had far-reaching and enduring effects on public health.

The conflict between the Western and Indian systems of medicine provided the colonial government with an opportunity to suppress the traditional systems of medicine in India. However, many people such as Hakim Ajmal Khan lobbied extensively against official suppressions in the early years of this century (Metcalf 1986). In 1918 the thirty-third meeting of the Congress (party) resolved, in recognition of the comparatively wide prevalence of the Ayurvedic and Unani systems of medicine in India and their undeniable claims to usefulness, to strongly recommend to the Government of India the desirability of taking definite steps to secure to them the advantages accorded to the Western system under the then government policy. However, modernizers such as Nehru supported the spread of Western scientific medicine, and Western-style doctors also joined Congress. Congress and non-Congress parties never committed themselves exclusively to one side or the other, as became clear during the 1920s (Jeffrey 1988).

Notwithstanding the fact that India achieved independence more than four decades ago, the gap between regionally developed and backward areas has actually widened. Political economic forces play a dominant role in the shaping of communities' health services, through decisions on resource allocation, personnel policy, choice of technology and the degree to which the health services are to be available and accessible to the population. In addition, whenever there are cuts or readjustments in plan commitments, the health sector is generally among the first victims. Expediency rather than any well-defined principles appears to govern the setting of priorities in the planning process, the allocation of funds, the allotment of portfolios and the appointment and transfer of personnel.

India's Ministry of Health and Family Planning instituted a Group on Medical Education and Support Manpower which, in a 1975 report, rejected the Western model of health services as characterized by huge costs, over-professionalization and loss of individual autonomy. The Group suggested an alternative model which would be based on communities' responsibilities in health and involve every individual. It recommended that this alternative model should integrate professional services with a new cadre of semi-professionals recruited from among the community itself, and that health services should be reoriented towards prevention rather than cure. Unique features of this post-1975 rural health scheme are:

1 Community participation in the health care delivery system, as reflected in locally chosen community health workers (CHWs).
2 Attempts have been made to overcome the limitations of health services

inadequately delivered through government-controlled professional staff alone, by supplementing them with a band of trained and trusted voluntary workers.

3 For the first time, the allopathic and Indian systems of medicine (Ayurveda, Unani and Siddha, including homoeopathy) are being combined to provide economically feasible and culturally acceptable 'appropriate technologies' in the rural areas (see also Chapter 4).

Table 14.1 Establishment of primary health centres and sub-centres in India since the First Plan

	PHCs	Sub-centres
First Plan (1951/2–1955/6)	67	
Second Plan	2,565	
Third Plan	4,919	22,826
Fourth Five-Year Plan	5,283	33,509
Fifth Five-Year Plan	5,484	47,112
Sixth Five-Year Plan	7,284	82,946
Seventh Five-Year Plan (1985–90):		
by 1987	14,145	98,987
by 1989–90	23,000	137,000

Source: Ministry of Health and Family Welfare (1987, 1990)

REGIONAL DISPARITIES IN HEALTH SERVICES IN INDIA

The establishment of primary health centres

Various components of India's health services have increased greatly in numerical terms. In the First Plan of the early 1950s, there were only 67 primary health centres (PHCs) in the country (Table 14.1). This rose to 14,145 by the middle of 1987 and to 23,000 in 1989–90. Similarly there has been substantial increase in the number of beds, doctors and nurses over the past forty years, and the availability of doctors, nurses and beds is now close to desired norms set earlier. The availability of doctors, nurses and beds has increased considerably during the period 1978–87. However, such increases in the number of health facilities are tending to create greater disparities between developed and underdeveloped areas of the country.

Per capita expenditure on health

A major problem is how to achieve balanced and uniform health care planning in the face of uneven expenditure on health in different states of India. There are large variations in per capita expenditure on health and family welfare among them. For example, during the period 1975–6, Uttar

219

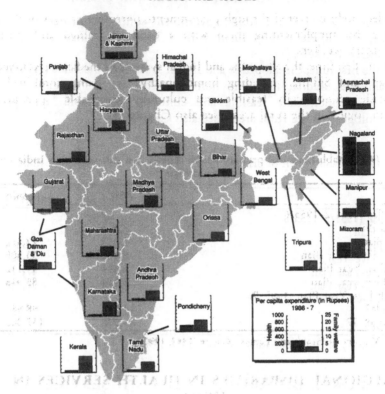

Figure 14.1 Per capita expenditure on health and family welfare in India 1986–7

Pradesh and Bihar states, where about a quarter of the country's population live, spent only about Rs 5 per person per annum. The per capita expenditure in these states rose to Rs 21 and Rs 15 respectively by 1983–4. Significantly, these areas also experience several types of communicable diseases, and, with the exception of certain districts of western Uttar Pradesh and southern Bihar, the area is backward both agriculturally and industrially. By contrast, most states of the north-eastern region and the southern part of the country enjoy relatively high per capita expenditure on health. For example, in 1975–6 per capita expenditure on health in the north-eastern region (Sikkim, Assam, Meghalaya, Manipur, Tripura, Nagaland and Arunachal Pradesh) ranged between Rs 38 and Rs 49, with the highest per capita expenditure in Nagaland of Rs 120, rising to 281 in 1983–4. Per capita expenditure on health is also high in the southern region. Pondicherry and Goa, and Daman and Diu states spent Rs 66 and Rs 65 in 1975–6 respectively per person per annum and Rs 170 and Rs 246 per person in 1986–7. Imbalances in services inevitably occur in the face of such wide disparities in expenditure on the health sector in different regions. Similar variations have remained in the late 1980s, as shown in Figure 14.1.

The distribution of hospital facilities

Wide spatial variations exist at the national level in the pattern of hospital facilities in India, and these have persisted since the 1950s. An index of hospital facilities in relation to population ranges between 5 and 35 and more hospitals per million of population. In 1957 in areas of general high population density, such as the Gangetic plains and the west of the country, adequate hospital facilities were lacking in comparison to low population density areas. With the exception of tourist resorts such as Simla, Naini Tal and Darjeeling, most of the mountainous areas had inadequate hospital facilities. Similarly, the tribal areas of Bastar, Koraput and Araku Valley had only 5 to 10 hospitals per million of population. By contrast, some less populated areas such as eastern Rajasthan and western Madhya Pradesh saw the index of hospital availability rise to 35.

However, the distribution of hospital facilities highlights imbalance in the growth of population and the growth of hospitals between the mid-1950s and mid-1970s. The growth of hospitals was not coterminous with the increase in population in eastern Rajasthan and western Madhya Pradesh. Rather, areas of high hospital availability were scattered countrywide. The special preference accorded to the low population density north-eastern region and Ladak area of Jammu and Kashmir resulted in their having high availability of hospital facilities. Certain areas with economic and strategic significance, such as the military centre of Gorakhpur and mineral-rich and industrially developed areas of the eastern part of the country, also show comparatively high availability of hospital facilities. Facilities are also well provided in educationally and industrially developed Kerala State. Nevertheless, even by the mid-1970s, in most of the high population density areas of the north, central, western and southern regions, hospital facilities remained utterly inadequate and well below the national average, although the process of regional development had touched almost all areas. By 1983 the higher concentration of hospital facilities lay mainly in the north-eastern and southern areas. The backward areas in the northern plains and in central Madhya Pradesh have not benefited much from the process of health care development of the last three decades.

Distribution of beds

Availability of beds (as opposed to institutions) is an important indicator of health services and it, too, is very uneven. In the mid-1950s, most districts in the north and the south possessed fewer than 25 beds per 100,000 of population. Districts with bed facilities ranging between 50 and 150 per 100,000 of population were widely scattered. The northern areas of Himachal Pradesh and Punjab, some districts in western Rajasthan, Gujarat, western Madhya Pradesh, western Maharashtra and the western coastal areas show

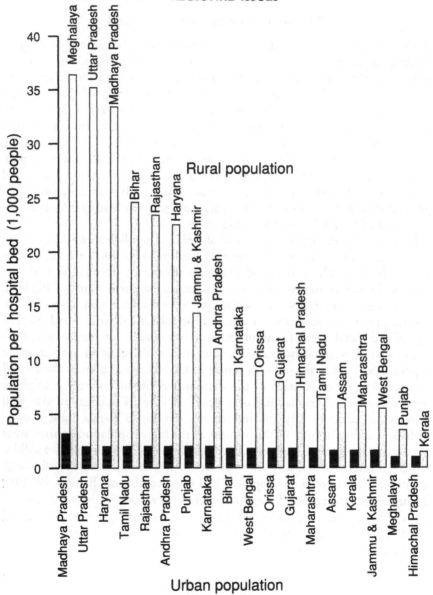

Figure 14.2 India: population per hospital bed, urban and rural, by states
Source: Based on *Ministry of Health and Family Welfare* (1987: 139)

high availability of beds in comparison to the high-population-density areas of the north-central and southern parts of the country. By 1975, with the exception of some districts such as Gorakhpur and hilly districts of Uttar Pradesh, growth in the availability of beds has been in the already-developed

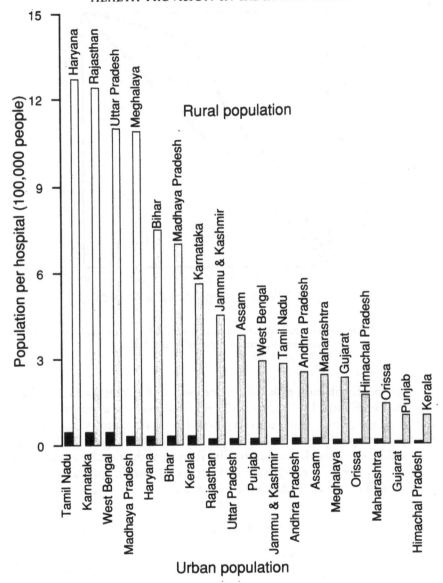

Figure 14.3 India: population per hospital, urban and rural, by states
Source: Based on *Ministry of Health and Family Welfare* (1987: 139)

areas. Again, by 1983 some parts of northern and central regions show 25 beds and fewer per 100,000 of population. The western states of Rajasthan and Gujarat and eastern coastal areas also fall into the lowest category of provision. Increases by 1983 in the bed-to-population ratios are evident in Kerala, West Bengal, Punjab, Haryana and Delhi. A comparative picture

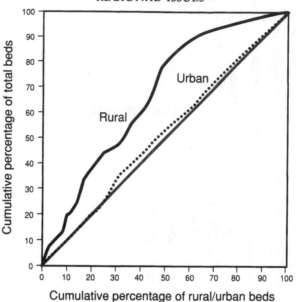

Figure 14.4 India: Lorenz curves for urban and rural health care facilities, based on the distribution of hospital beds (1985)

showing the distribution of hospitals and beds in different states in India is presented in Figures 14.2 and 14.3. These illustrations are self-explanatory and reveal the grim regional disparity prevalent in the country, on the basis of population per hospital facility or per bed. Lorenz curves for 1985 show the considerable irregularities in the distribution of the health facilities both in urban and in rural areas (Figure 14.4). It is clear that inequalities are more widespread in rural areas than in urban areas.

Doctor-to-population ratios

There has been an increase in ratios of population per doctor in general between the mid-1970s and mid-1980s. This has varied from *nil* or little difference in the most populous states of Bihar and Uttar Pradesh to a substantial increase in the ratio in Rajasthan, from about 4,000 of population per doctor in 1977 to 11,000 in 1986, although this did decline to 3,461 in 1990. Substantial increases have also been noticed in respect of Delhi and Punjab. There is a clear indication that, in spite of the high production of physicians in India, doctors are leaving government jobs and either emigrating, joining the private sector or entering private practice. In 1990, doctor-to-population ratios ranged from about 15,631 in Uttar Pradesh to 820 in Chandigarh, indicating the huge disparities in this index.

224

Table 14.2 Distribution of villages/urban blocks in India and estimated population by type of medical facility and distance

	Villages/urban blocks having medical facility (%)	Population having medical facility (%)
Rural:		
Within 2 km	40.9	35.3
2–5 km	23.7	24.9
5–10 km	20.5	23.7
10 km +	14.9	16.1
Urban:		
Within 2 km	84.4	83.9
2–5 km	13.0	13.6
5 km +	2.6	2.5
Total:		
Within 2 km	56.5	44.1
2–5 km	19.8	22.8
5–10 km	13.5	19.7
10 km +	10.1	13.4

Source: Office of the Registrar General (Census), New Delhi

Distance and health care

An important aspect in the study of health care is the distance between settlements and health facilities. A study by the Registrar General's Office in connection with infant and child mortality revealed that only about 41 per cent of villages and about 35 per cent of the total rural population of the country have access to a health facility within 2 km. On the other hand, health facilities are available within 2 km for about 84 per cent of the total urban population of the country (Table 14.2). Such urban–rural disparities in accessibility are, of course, common in many countries.

Growth in population and health infrastructure

Facility increases have actually outstripped population growth, but from a low base. For example, population grew by 25 per cent between 1957 and 1975, but the number of hospitals, beds and doctors rose by 190 per cent, 230 per cent and 84 per cent respectively. There are, however, variations in the growth of bed and hospital facilities in urban areas. For example, among towns, in growth of the number of beds, Bombay recorded the highest increases (137 per cent), and the lowest in Calcutta (44 per cent). The growth in hospital facilities per 100,000 people varied between 5 per cent in Calcutta and 27 per cent in Madras. Wide rural–urban imbalances remain in the distribution of health facilities. Rural areas, with about 80 per cent of the total population, have under one-third of the total number of hospitals, one-seventh of total beds and well under half the total doctors in the country.

Disparities in the provision of specialized hospitals

There is not only unequal distribution of beds, general hospitals and doctors, but also uneven distribution of sophisticated hospitals and medical research centres nationally, as is typical in many Third World countries. For example, the All-India Institute of Medical Sciences and Safdarjang Hospital are located opposite to each other in New Delhi, and postgraduate institutes were opened in Chandigarh and Pondicherry as prime institutes because of political jealousy of the pre-eminence of Delhi's institute. Recently, Tamil Nadu has also established a further medical university in Madras city, a place already endowed with many excellent medical centres. It has therefore been reasonably suggested that such new medical universities should be allocated to some other district centre in the southern part of the state, to boost well-equipped medical training institutes where the need is great. Nevertheless, in February 1990, the health minister of the state of Tamil Nadu laid the foundation for a new hospital in Madras city, paying lip-service to the need to establish such hospitals in rural areas. There is great scepticism concerning the existence of political will to provide health care and development in rural areas.

HEALTH AND HEALTH CARE: EXAMPLES FROM ZAMBIA

Health conditions in Zambia (formerly Northern Rhodesia) were historically poor, with malaria as the dominant cause of morbidity and mortality until the 1960s. There was even discrimination in the distribution of quinine in order to combat malaria during the colonial period, particularly in the early years of this century. Gelfand (1961) pointed out that in Northern Rhodesia quinine was issued free to all officials and to some Africans; because of malaria, Sir Harry Johnston declared it could never be a 'white man's country'. At the same time, Gelfand suggests, malaria and Africa were almost synonymous, with very little that physicians could do to ameliorate the disease. Missionaries were well aware of malaria, and both the Universities' Missions to Central Africa and the London Missionary Society often spoke of the sacrifices made by missionaries when, as one of their workers was cut down by fever, another replaced his fallen comrade. In 1907, the death rate in the country at that time for the European population was given as 114 per 1,000 (Gelfand 1961).

The major causes of morbidity and mortality in Zambia in the 1980s were diseases that can generally be controlled through preventive measures. These measures include provision of adequate water supply, proper sanitation, public health education and vector and rodent control, as well as the availability of health facilities. Therefore, an expensive urban-based Western medical system which relies on treatment rather than on prevention is unlikely to be an effective way to tackle such a country's health problems. Nevertheless, life expectancy at birth has increased from 40 years in 1964 to 48 in 1979 and to 54 years in 1990; death rates for infants and children have also been reduced.

Medical services by missionaries

During the colonial period, different health facilities tended to exist for Africans and Europeans, and better official health facilities were concentrated in urban and mining centres where most Europeans lived. By contrast, missionaries overwhelmingly provided health care in rural areas, and curative rather than preventive medicine was fostered. Even after independence in 1964, inequalities in health care provision continued to widen, and curative rather than preventive medicine continued to dominate. Nevertheless, concern during the colonial period is revealed about the health of the people; and during the 1940s there is clear evidence of an official desire for doctors to be trained in public health and nutrition.

No discussion on health service patterns in Zambia will be complete unless the historical impact of missionary medical personnel is clearly understood. Missionaries were the first Western physicians to arrive in Zambia. Mission work in the country started in the 1880s with the Paris Missionary Society in Bulozi and the London Missionary Society in the area between Lake Tanganyika and Lake Mweru in the northernmost part of Zambia. Both societies had medical missionaries on their staff. Several other missionary societies of many denominations brought physicians with them prior to 1900.

The significance of the medical work by various missions lies in the fact that they were mainly confined to the rural areas, and a large number of mission hospitals are still located in rural areas. Another important feature is that mission health services were available irrespective of race and economic status. This was why the London Missionary Society had turned down in 1899 an offer of government financial assistance 'to attend to Europeans to the neglect of their missionary duties' (i.e. the serving of the local African population).

However, the colonial administration in the early twentieth century was able to reduce mortality levels by introducing measures which gradually but effectively reduced mortality from tropical diseases, especially malaria. By 1924, every important station, post or district had its own government doctor, and in all the larger towns European nurses were placed in the hospitals, which were able to provide adequate and efficient care for the patients. Hygiene specialists also carried out effective propaganda on the dangers of malaria, the need to take quinine regularly as a prophylactic and the importance of clearing the bush and draining collections of water near the house.

The mission medical services with their main focus on the rural population helped set the typical patterns of health service distribution at the time of Zambia's independence in 1964. Of 81 mission institutions, 51 were located in rural areas. However, many colonial officials saw the medical work of missionaries as competitive or contradictory to the government's work, and no financial assistance was granted to these missions. In particular, if public finance to build hospitals were to be granted, the government in the 1940s wished to retain the right to engage and dismiss staff.

Health service patterns in Zambia today

The rapid population growth in Zambia as in many other developing countries has posed serious problems for health planners. Natural increase is now over 3 per cent per annum, implying that health care provision also has to increase at a similar rate merely to keep pace. On the one hand, the government is committed to a free and expanding medical service with health centres within walking distance in the rural areas; on the other, due to the severe financial problems that the country currently faces due to declining copper prices, it is very hard to maintain even the existing health services. Many medical personnel, often expatriates, have left since the devaluation of the local currency (kwacha), particularly since the early to mid-1980s. In addition, because of poor conditions of service, even Zambian doctors are leaving the country for Zimbabwe, Botswana and other countries of Southern Africa. Indeed, it has been estimated that there are only 500 doctors in the country as against the requisite 1,000. Several attempts to recruit doctors from abroad have failed. There are similar concerns over the scarcity of medicines in government hospitals and health centres. To try to counter this, the government has introduced health service charges. This may, however, have a deleterious effect on utilization, which has been noted in many instances where user charges or cost recovery attempts have been introduced.

The distribution of population, hospitals and rural health centres

Population distribution in Zambia is very uneven; there are pockets of high population density along the railway line, particularly in the mining Copperbelt Province and the Lusaka area. Elsewhere, the population is sparsely distributed in the northern, north-western and south-western parts of Zambia. In terms of the distribution of health care facilities, most hospitals are located along the line of rail, mainly in the Copperbelt Province and in the Southern Province. Provinces such as the Northern and

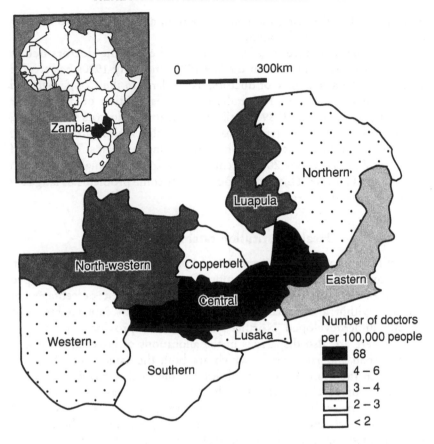

Figure 14.5 Zambia: doctors per 100,000 of population, by province, 1989

North-Western have fewer hospitals than needed. Rural health centres tend to be generally more evenly distributed throughout the country. However, some hospitals are not ideally situated in relation to rural health centres.

Health-services-to-population ratios

In 1981 there were over 800 health institutions in Zambia, but there were considerable regional variations in health centre-to-population ratios. It has already been noted that there is an acute shortage of health personnel, particularly physicians, in Zambia. The regional distribution of physicians is highly uneven, although there have been minor changes in ratios. The urbanized provinces of Lusaka and Copperbelt showed the lowest physician/population ratios. The explanation for this is that physicians are mainly concentrated in the urban areas along the line of the railway. This was the area originally developed during the colonial period, and it is the hub of both industrial and urban development.

However, the departure of a large number of expatriate doctors due to declining economic conditions in the late 1980s and early 1990s has created a wide gap in the availability of doctors, particularly in the Copperbelt and Southern provinces. Figure 14.5 indicates the distribution of doctors per 100,000 of population in 1989. Until 1988 expatriate doctors from India, Pakistan and Bangladesh tended to provide the major proportion of physicians in Zambia. This pattern has now changed; almost one-third of doctors are Zambians, but 20 per cent are Cuban and a small proportion Egyptians and Chinese. This can of course create some problems of language and culture in medical treatments.

Health institution catchment areas

The identification of the catchment area of a particular health facility is often useful in overall spatial health planning. Data indicate that the developed provinces of Lusaka and Copperbelt have the highest percentages of population within a distance of 12 km from a health care facility. The regionally underdeveloped Northern and Western provinces are poorly covered. There are also disparities in the allocations of government health funds to urban and rural areas, which are both the causes and effects of imbalances in the availability of facilities. Of the total recurrent budget of the Ministry of Health in 1979, only 38 per cent was allowed to the provinces. Of the capital budget, the share of the provinces was less than 37 per cent. Three central hospitals, allocated along the line of the railway, alone accounted for about one-third of the total recurrent budget.

It has been recognized that the emphasis on providing services by way of hospitals, clinics and health centres, and relying largely on curative medicine, exacerbated inequalities. The government has thus accepted a primary health care programme with emphasis on rural areas. The package includes all the major components of PHC: health education, promotion of adequate nutrition and food supply, promotion and maintenance of safe water supply, basic sanitation, maternal and child services, immunization, prevention and control of locally endemic diseases, promotion of mental health and treatment of common diseases and injuries. Villages in rural areas and 'sections' in urban areas will be the basic units (focal points) for providing the above facilities through community health workers (CHWs) selected locally. The organizational structure is an interesting feature of the primary health care unit programme, and it has a geographic dimension. Under this programme, health zones will be created within each district. Each zone will have a headquarters with responsibilities for provision of health services to people living outside PHC units; for training, supervision and back-up of CHWs in villages outside PHC units; for promotion, supervision and evaluation of all PHC activities within the zone; for maintenance and distribution of drugs and other supplies to rural health centres within the zone; and to serve as a referral centre for patients from within the zone.

The zone headquarters will be a health establishment with adequate staff and transport. In one zone in each district, the district hospital will be the headquarters. In other zones, the headquarters will be an upgraded rural health centre. It is intended that the boundaries of a health zone should be such that all areas within the zone are accessible by transport from the headquarters. An area of 50 km radius around the headquarters is proposed, which is about two hours' drive on bush roads. There would normally be two or three health zones in each district of 20,000 to 50,000 of population.

The primary health care programme appears ambitious, and its success will depend upon the availability of funds and skilled personnel. As noted, there is an acute shortage of health personnel, especially physicians, in Zambia, and financial resoures are so meagre that it is difficult to assess the current success of this programme, which has been severely hampered by the non-availability of required health personnel. As in many other Third World countries, Zambia's primary health care programme is suffering because of lack of recurrent expenditure, supplies, transport and staff. In many rural areas, health centres have been desperately short of or totally without such things. Villages have not been visited by health staff for long periods because of lack of money to buy petrol. The shortage of spare parts for vehicles and the government's inability to buy them due to lack of foreign exchange are a major cause of transport problems. This in turn leads to a lack of field visits and supervision.

CONCLUSION AND SUGGESTIONS

The Zambian case has illustrated the existence of inequalities in the provision of health care in a typical large Third World country. As in India, inequalities have been widening over the years, despite many-fold increases in the health budget in every plan period. The chapter highlights the importance of politics and political will in health care in both India and Zambia.

Present health care patterns are based on a professional model with more concern for the quality of urban-based health care than for rural services or preventive care. Under this model, the role of indigenous resources is rarely accepted. The Kerala example in India showed allopathic doctors antagonistic to government plans for integration, and the establishment of private medical colleges to boost the production of doctors.

What countries in the Third World really need is a model of health care in which both personnel and hospitals and health centres are distributed in better accord with population distribution, with particular emphasis on rural, backward, tribal, mountainous and endemic areas. This demands the introduction of spatial planning at the micro-level in order to distribute facilities on the basis of equity. However, one of the serious problems in the health planning process is lack of understanding of the organization of regional structures and how to adopt equity and efficiency models in the provision of health facilities. Both equity considerations (to minimize distance) and efficiency considerations (to minimize aggregate travel) are in theory adopted in the formulation of many health care plans. Equity considerations aim to improve and expand health care to cover every settlement within a few kilometres of a health facility. Efficiency aspects concern the utilization of existing personnel and facilities as efficiently as possible. It is appropriate to study the socio-economic and physical framework of particular regions before emphasizing the relative importance of these two criteria.

An equity criterion is often suggested in the provision of amenities including health facilities. However, it is often assumed that the region where facilities are to be located is homogeneous. In addition, an efficient system of transport and communication with little variation, and homogeneity in the size and distribution of settlements, are also assumed. However, India is a vast country with wide socio-economic and physical variation, and Zambia is smaller in comparison (although still large). India is a large exporter of medical personnel, and Zambia a large importer. Therefore, no single model can be applied throughout both countries. The prerequisite for application of either an equity or an efficiency model is the demarcation of the area into homogeneous regions. These may be as follows:

1 Regions with high levels of socio-economic development. The application of an efficiency model would probably be appropriate.
2 Areas with special needs: backward, tribal, inaccessible, underdeveloped, hilly and mountainous areas. An equity model gives a way of providing health care to all segments of population.
3 Specific disease zones (for example, leprosy and filariasis). Such regions may need special health care or provision. Since socio-economic development is often low in such areas, an equity model is probably appropriate for their health care provision.

With regard to countries such as India, it is ironic that there is currently large-scale unemployment among doctors, but at the same time there are large numbers of hospitals and health centres in rural areas without medical staff. Many physicians are apparently unwilling to serve in rural areas, especially remote parts or areas with poor infrastructure. Therefore, simply to increase the number of doctors by itself would not solve the problem. Against this background, it would perhaps be suitable for India to adopt a model in which a large number of trained health assistants serve as a link between the people and the limited number of doctors in rural areas. This model has been adopted successfully in countries such as Vietnam and Ethiopia. There is no denying the fact that medical education and the skills of trained personnel need to be transformed, especially in Third World countries.

REFERENCES

Akhtar, R. (1991) (ed.) *Health Care Patterns and Planning in Developing Countries*, New York: Greenwood Press.

Banerji, D. (1979) 'The place of indigenous and Western systems of medicine in the health services of India', *International Journal of Health Services* 9: 511–19.

de Figueiredo, I.M. (1984). 'Ayurvedic medicine in Goa according to European sources in the sixteenth and seventeenth centuries', *Bulletin of the History of Medicine* 58: 227.

Gelfand, M. (1991) *Northern Rhodesia in the Days of the Charter*, Oxford: Blackwell.

Jeffrey, R. (1988) 'Doctors and Congress: the role of medical men and medical politics in Indian nationalism', in M. Shepperdson and C. Simmons (eds) *The Indian National Congress and the Political Economy of India 1885–1985*, London: Avebury.

—— (1991) 'The impact of socioeconomic and political factors on the provision of health care in India' in R. Akhtar (ed.) *Health Care Patterns and Planning in Developing Countries*, New York: Greenwood Press, pp. 99–119.

Kalapula, E.S. (1991) 'Health care delivery patterns and planning in Zambia: colonial, populist and crisis intervention approaches', in R. Akhtar (ed.) *Health Care Patterns and Planning in Developing Countries*, New York: Greenwood Press, pp. 199–225.

Klein, I. (1972) 'Malaria and mortality in Bengal, 1840–1921', *Indian Economic and Social History* 9: 132–60.

Metcalf, B.D. (1986) 'Hakim Ajmel Khan: Rais of Delhi and Muslim leader', in R.E. Frykenberg (ed.) *Delhi through the Ages: Essays in Urban History, Culture and Society*, Delhi: Oxford University Press.

Ministry of Health and Family Welfare (1987, 1990) *Health Information of India*, New Delhi: Central Bureau of Health Intelligence, Directorate of Health Sciences, Ministry of Health and Family Welfare.

Ramasubban, R. (1978) *Health Care for the People: The Empirics of the New Rural Health Scheme*, Technical report, Lucknow: Giri Institute of Development Studies.

—— (1982) *Public Health and Medical Research in India: Their Origin under the Impact of British Colonial Policy*, SAREC Report No. R4, Göteborg: Swedish Agency for Research Co-operation with Developing Countries.

15

HEALTH CARE IN LATIN AMERICA

Susana Isabel Curto de Casas

This chapter attempts a review of health and health care in Latin America. This is a huge and diverse region, stretching from the Mexico–United States border to Tierra del Fuego, and including the Caribbean. Some 450 million people live here. Countries and conditions differ, but in this chapter an overview of fundamental issues and common health themes is given. In particular, emphasis is placed on the national and international settings within which health and health care systems evolve. The development of health care programmes and social security systems in Latin America has been very much determined by changing economic, social and political conditions in the region. The majority of countries have evolved capitalist systems which depended on the availability of land for labour-intensive practices oriented towards the supply of primary products for export.

Since 1880 Latin American nations and the United States have attended various Pan-American conferences to establish uniform quarantine regulations in an effort to remove barriers to sea communication. Indeed, the national health departments of Argentina (1880), Uruguay (1895), Brazil (1897) and Paraguay (1899) were largely organized around the issue of health and maritime sanitation. The United States' geopolitical interests in the area were linked to the creation of the Pan-American Sanitary Bureau (1902) and the national health departments in Cuba (1902), Haiti (1919), Panama (1926) and the Dominican Republic (1920) during the US military occupation of these countries. The initial mission of these departments was to combat the infectious diseases that hampered maritime trade. Their primary emphasis was on the fight against the contagious diseases in the major port cities which justified quarantine, such as yellow fever, cholera, bubonic plague and smallpox. As a result of these sanitary efforts, yellow fever disappeared from Havana, Panama, Guayaquil and Rio de Janeiro in the early years of the twentieth century.

In subsequent periods, the sanitation responsibilities and activities of health departments increased to deal with problems affecting areas of agricultural production. The initial land sanitation programmes were centred on the control of hookworm disease and malaria in order to combat the low productivity of workers in export-producing regions. According to the reports of the first Pan-American conferences, the objective of these programmes was to make the land fit for agriculture; so again, economic underpinnings were crucial.

National health departments were established in Honduras (1917), El Salvador (1920), Costa Rica (1922), Guatemala (1925) and Nicaragua (1925). These were organized around hookworm programmes carried out in banana- and coffee-producing regions with the support of the International Health Commission of the Rockefeller Foundation. Maritime sanitation activities were the responsibility of foreign investors such as the United Fruit Company.

This health care approach, based on the fight against infectious diseases by sanitary engineering and vaccination, reached its highest point in 1950 with the introduction of DDT, which widely reduced endemic areas of malaria and yellow fever. This approach decreased the general mortality rates to 17 per 1,000, but infant mortality rates remained high (119 per 1,000), especially in tropical regions where they were around 135 per 1,000 (PAHO 1982). During the 1970s, many agencies, including the United States Agency for International Development (USAID), the United Nations Children's Fund (UNICEF), the Inter-American Development Bank (IADB), the Canadian Public Health Association and the Rotary Club, supported extensive infant immunization campaigns such as the Expanded Programme on Immunization (EPI) in the late 1970s to bring to all children protective immunization against six diseases (diphtheria, whooping cough, tetanus, measles, poliomyelitis and tuberculosis) and oral rehydratation therapy (a glucose–salt solution in pre-packed sachets) in the 1980s (see Chapter 9). Together, these disease-oriented programmes expanded life expectancy to 66.6 years. Today, life expectancy is over 70 years for about half of Latin America's population.

In part as a consequence of this demographic change, chronic disease rates grew and now account for up to 60 per cent of all deaths in the Southern Cone of the continent, 57 per cent in the Caribbean, 45 per cent in tropical South America and 28 per cent in Central America, including Panama and Mexico. As a result, new problems have been added to the old infectious disease–sanitary model which have been magnified by the contradictions of regional post-war economic growth. Many of the middle-income countries have a type of epidemiological transition in which they experience the problems of both the developed and the developing world.

CONTRADICTIONS OF POST-WAR ECONOMIC GROWTH

Since 1940, Latin America has undergone an important socio-economic transformation. Its gross national product (GNP) grew by 5 per cent in the 1950s, by 5.5 per cent in the 1960s and by 6.3 per cent in the 1970s. Its international trade and productive capacity also expanded. These conditions provided the illusion that development was on the right track; but new problems were being generated, given the high and increasing population growth rates. As a result, the rate of GNP growth per capita was reduced to 2.6 per cent, and unemployment and poverty have been steadily increasing.

This economic growth was largely based on an import-substitution industrialization model. As a consequence, industry started to grow, to diversify and to expand. Between 1950 and 1980 the industrial share of GNP grew from 17 per cent to 25 per cent. Agriculture was modernized, and new technologies and agro-industrial activities transformed many traditional systems. However, the introduction of industrial farming and the utilization of cereal crops to feed cattle diminished human food production. So modern agriculture was unable to satisfy the higher food demands of demographic growth; therefore, in 1960, food production per capita was less than in pre-war years.

In addition, unemployment was an unexpected consequence of modernization; agricultural underemployment and unemployment grew from 13 per cent in 1950 to 17 per cent in 1970 and 19 per cent in 1980. Labour which was made redundant by industrial and agricultural modernization was displaced to the urban service sector, which grew faster than other productive activities. Despite this, a large proportion of the population was unemployed, especially in the big cities where unemployment rates, as calculated by the Economic Committee for Latin America and the Caribbean (CEPAL), started to grow from 6.5 per cent in 1970 to 11 per cent in 1985.

Unemployment has increased regional poverty. At the beginning of the 1970s, 40 per cent of Latin America's population lived in poverty (lacking income to satisfy minimum necessities) and 20 per cent were indigent (lacking income to have food-minimum expenses). The problem was worst in rural areas, where 62 per cent of the population lived in poverty and 34 per cent in indigence. In cities, 26 per cent of the population lived in poverty and 10 per cent in indigence.

Since the 1970s, conditions have not improved. In 1980 the rates grew to 70 per cent for poverty conditions and 40 per cent for indigence. In 1985 the Regional Program of Employment for Latin America and the Caribbean (PREALC) estimated the poverty level at 30 per cent of Latin American families. In some countries such as El Salvador, Guatemala, Haiti and Nicaragua, poverty rates grew to more than 50 per cent (PAHO 1990a). This has had important health implications. For example, as a consequence of the increase in regional poverty, the availability of calories per capita in some countries such as Chile, Haiti, Peru and Uruguay is below 1965 levels. According to the World Bank, in 1989 only Argentina and Mexico have intakes of above 3,000 calories per day, while Bolivia's does not amount to 2,000.

Modernization has produced an acceleration of the growth in the largest cities. Latin America's population has become predominantly urban since 1960 and that of the Caribbean zone since 1970. Excessive urban growth rates resulting from rural–urban migration and from high natural increase have created chaotic and disorganized patterns which have contributed to deterioration of the environment and of quality of life. Infrastructure and services cannot meet demand, and the housing deficit has led to segregation and marginalization for large segments of the population. Between 20 and 30 per cent of the population of large cities in Latin America live in marginal areas; 21 per cent are not connected to a system for drinkable water; and 52 per cent are not connected to a sewerage system. Garbage accumulates in streets or on waste ground. When it burns or percolates into the groundwater supplies, it becomes an agent of air or water pollution and a major threat to public health. Industrial and domestic waste water are rarely purified, and they often mix with organic compounds and heavy metals flowing into subterranean water and rivers, which provide drinking water. As a consequence, gastro-enteritis and diarrhoeal diseases are among the ten leading causes of death.

Violence is a relatively new urban phenomenon, influenced by cultural, political, socio-economic and familial factors. Alcohol, tobacco and drug consumption is increasing among youths and adults. For example, cocaine is estimated to be consumed by between 4 and 16.8 per cent of secondary-school students and by between 1.4 and 6.7 per cent of the 12–45-year-old group in some neighbourhoods and marginal areas. The inhalation of paints, glues and organic and industrial solvents is also growing among children and pre-adolescents. It is estimated to involve 4.4 per cent of secondary-school attenders in Mexico and 3 per cent in Chile (PAHO 1990a).

Accidents are the principal cause of death in the 5–44 age group in twenty Latin American countries. The rates range from 3.3 per cent of all deaths in Jamaica to 15.6 per cent in El Salvador and 17.2 per cent in Honduras. Homicide is the leading cause of death for the 15–24-year-old group in Colombia, and suicide is the major single cause of death in young women in El Salvador and Trinidad and Tobago (PAHO 1990a).

The other aspect of high urban concentration in Latin America is the low density of rural population. Only 28 per cent of Latin America's population live in rural areas, and they have poor health care services: 45 per cent of the rural population do not have drinkable water, and 68 per cent do not have access to a sewerage supply.

Rural migration between bordering countries is increasing. Many migrants are war or political refugees or temporal workers related to agricultural frontier colonization opportunities. The movement of workers will increase in the future with the economic integration of countries such as Argentina, Brazil, Paraguay and Uruguay in the MERCOSUR ('South Common Market') free trade area. Rural–rural migration movements can produce two main sets of health care problems: first, the need to give health care to illegal workers and to improve the precarious social condition of workers; and secondly, the need to solve the difficulties of sanitary control in border areas.

MODERNIZATION, SOCIAL CHANGES AND HEALTH CARE

Latin American health care systems are the product of struggles between social groups, some to obtain new privileges and others to maintain the privileges obtained. Moreover, health care and social security systems may be regarded as social control mechanisms put into practice by the political powers in individual countries.

An urban proletariat and a working 'aristocracy' of service workers were the new social groups consolidated by the modernization and urbanization processes. These groups had great political influence because they had the power of labour unions. They had industrialized societies as a reference against which to measure education and access to social services. Hence, in many countries, children were incorporated into primary education, and universities had a massive student enrolment but had remarkable inefficiency. Education was also a means of social mobility for a wide range of marginalized groups in the population. However, illiteracy has continued at a high rate in rural populations, especially among native groups in the continent.

In some pioneer countries such as Argentina, Chile, Uruguay, Brazil and Cuba, which have had social services and health care services since the 1920s, access was restricted to certain power groups. Accompanying the processes of urbanization and education, labour unions and political mobilization were advanced. As a consequence, previously unprovided-for groups gained access to health care programmes through their new political power. These Latin American health care systems were considerably influenced by the populist political movements which flourished in the 1940s in the region. Populism in Latin America represented a multi-class, nationalist political movement that enjoyed mass support among the lower classes and espoused apparently anti-establishment policies, but was in fact organized and led by ruling-class politicians. The promotion of these social security programmes responded to the needs of industrialization and served as an important mechanism of social stabilization. Populist governments such as that of Getulio Vargas in Brazil and Juan Perón in Argentina started a process of state redefinition in health and social assistance. As a result, the state responded to the social demands of the times.

During the 1940s and 1950s the pressure of wage-earning workers on the protectionist state ensured growth in the coverage of health care and social services. This resulted in very liberal loans which promoted imbalanced conditions and a financial disorder of the systems. Subsequently, national and international technical studies recommended the unification of health care systems and elimination of expensive privileges, but power groups forced the delay of reforms for decades. The reforms could be implemented only when political changes reduced the power of the pressure groups during the 1960s and 1970s. In Cuba, for example, the system was unified through state control; other countries such as Argentina and Uruguay created central bureaux of co-ordination, and Chile created a new system based on privatization.

On the other hand, countries which established their social security systems in the 1940s, such as Mexico, Colombia, Venezuela and Peru, were influenced by International Labour Organization trends and the British Beveridge Report in creating general systems which covered all of the population (universal coverage). One of the advantages of this type of system was that it prevented or minimized stratification problems in societies with low industrial development and predominantly rural populations. Another advantage was that it did not have the problems characteristic of the old systems of pioneer countries. These may be called unified systems. However, there is a great problem because these systems have usually been restricted to the most important cities only. In the Spanish-speaking Caribbean, excluding Costa Rica, Panama and Cuba, social security and health care systems appeared in the 1950s and 1960s. The English-speaking Caribbean has a health care system based largely on the influence of the British system.

In the 1980s Latin American societies almost universally experienced a new process of change influenced by world tendencies of democratization and decentralization. The importance given to individual action ended state expansion and has largely changed the health care models. The new model tends to be for local health systems (*sistemas locales de salud*: SILOS), characterized by community participation in planning, implementation and evaluation of activities. SILOS are usually based on a geographically defined population. Their objectives are to care for individuals and communities, to administer health resources and to facilitate social participation (PAHO 1990b).

SILOS are part of the present strategy of the Pan-American Health Organization (PAHO) to reorganize health care using the primary health care system (PAHO 1988a). They could conceivably generate modern health care systems for an élite and a growing middle class, and primary health care level services for the poor. The change in the health care model will be more difficult to achieve in countries with large indigenous populations or with substantial ethnic, religious or economic differences between social classes. Community participation cannot be developed without protection from the system, and of course, as discussed in Chapter 12, it can be very difficult to sustain. As a consequence, the system needs to be flexible, should permit the community to assume health

Table 15.1 Internal contradictions between national health services and community participation

National health services	Local health services
Curative medicine	Preventive medicine
Physician–patient relationship	Community–environment relationship
Complementary and controlled community participation	Community participation in execution and evaluation of activities
Priorities given by professionals	Priorities considered by community

responsibilities and at the same time should supply theoretical and practical knowledge for population health needs.

Deconcentration and decentralization are independent processes in the region. Decentralization can be achieved only through political will or decision, because of the persistence of the Spanish tradition of administrative centralization based on the premise of national unification. This tradition produced powerful centralized administrations which can supervise the decentralization process, because they use deconcentrated territorial or administrative units. Deconcentration can be undertaken within decentralization and the delegation of authority and responsibility (PAHO 1990b). Sometimes the transfer of services and authority to smaller administrative units is not accompanied by the transfer of money. This may in effect negate the devolution of power or responsibility. There are also many internal contradictions between national health services and community participation (Table 15.1). Some of these have been touched on in Chapter 12.

HEALTH CARE SYSTEMS

In Latin America there are different, parallel and coexisting health care systems which can interact. The major ones are:

1 Private health care systems with profit-making characteristics, for high-socio-economic status groups. They are financed by direct payment in advance and private insurance.
2 Health care systems provided by health department or health ministries, financed by the state through taxes. They are usually free and theoretically are oriented to people who are not covered by social security, including the indigenous population.
3 Social security health care designed for workers and their families. They are financed by three sources of contributions: from the state, from workers and from employers (sometimes the state is also the employer). The contribution is based on salary levels. In most countries the coverage is limited. This model, based on the ideals of Bismarck's Sickness Insurance Act of late-nineteenth-century Germany, has not worked well

in Latin America because the majority of the labour force is composed of independent workers who cannot or do not contribute, or agricultural workers with low incomes.

There are different combinations of these health care systems:

1 Countries in which the responsibility for health care is the exclusive preserve, or almost so, of health ministries, such as in Cuba, Nicaragua, Haiti and the English-speaking Caribbean.
2 Countries in which the health care of the majority of the population is covered by the social security system: Argentina, Brazil, Costa Rica, Mexico, Panama, Uruguay, Venezuela.
3 Countries in which participation between the Ministry of Health and social security prevails: Bolivia, Colombia, Ecuador, El Salvador, Guatemala, Honduras, Paraguay, Peru and the Dominican Republic.

Health care coverage has recently been estimated as providing for 61 per cent of Latin America's population (Mesa Lago 1988), which is the highest coverage of regions in the underdeveloped world. However, it is enlarged by Brazil's rates, which are imprecise. If Brazil's rates are excluded, Latin American coverage rates fall to 43 per cent of the population. A few countries match the levels of developed countries (which run at 75–100 per cent): these include Cuba, Argentina, Costa Rica, Uruguay and Chile. The majority of countries have lower coverage: under 20 per cent, such as Paraguay, Peru, Guatemala and Colombia; and under 10 per cent, such as Nicaragua, Ecuador, the Dominican Republic, Honduras, El Salvador and Haiti.

In some countries, the financial stability of health services is precarious or in crisis, precipitated by a range of factors. These include increased unemployment and reduction of real salary levels because of inflation and company failures which reduce contributions to the system and taxes collected. Unemployed people do not contribute but receive health care, as do the poor in the state-financed systems. Inflation increases evasion of payments and delays in contributions and taxes. State contributions may also be diverted to other uses such as the South Atlantic War or the Peruvian Sierra Road construction. There is the pervasive Third World predominance of expensive curative medical care over often cheaper and more appropriate preventive medicine. Last but not least, there is a preference for high technology and excessive use of medicines. The financial crisis in health care has been tackled by drastic reforms and with less visible methods such as permitting services to deteriorate, which influences the medical attention received by lower-income people.

There are more than 1 million hospital beds, and the overall bed-to-population ratio is 2.5 beds per 1,000 of population. This rate has generally decreased, with the exception of Cuba, some Caribbean English-speaking

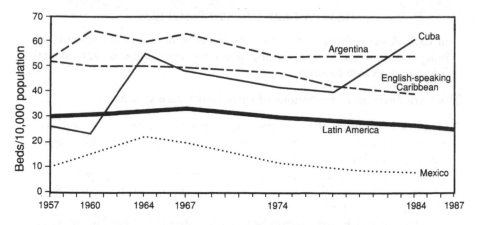

Figure 15.1 Latin America and the Caribbean: hospital beds per 10,000 of population, 1957–87

islands, Haiti and Brazil. The countries with high rates are Argentina (5.4 beds per 1,000), Uruguay (5.0), Cuba (6.1) and the English-speaking Caribbean. Lower rates are in Mexico (0.8 beds per 1,000), Honduras (0.9), Haiti (1.0), the Dominican Republic (1.2) and Paraguay (1.4) (PAHO 1988b) (Figure 15.1)

An important tendency since the early 1960s has been a growth of private hospitals and hospital beds. Andean countries, Central America and the English-speaking Caribbean have a predominance of public hospitals, while, on the other hand, Argentina, Brazil, Chile, Uruguay and Mexico have a predominance of small private hospitals and clinics. As a consequence, a few great public hospitals provide the major hospital capacity in the public sector. Overall, Latin America has 55 per cent public hospital beds and 45 per cent private hospital beds.

The number of physicians and nurses has grown since the late 1950s but at different rates in different countries. The higher ratios of physicians per capita are in the Southern Cone, where Argentina (27.0 physicians per 10,000 of population) exceeds the United States (21.4) and Canada (19.6). The highest ratios of nurses per capita are in the English-speaking Caribbean, Belize, Panama and Cuba. The growth of physicians is not related to the limited growth of the other medical resources. In Latin America as a whole, there are two physicians per nurse, and in the Southern Cone and Brazil there are five physicians per nurse. Only in the English-speaking Caribbean is the rate three nurses to one physician.

THE 1980s CRISIS AND ITS EFFECTS ON HEALTH CARE

In the 1970s Latin America had easy access to international credit. The growth of interest rates and the 30 per cent fall in international prices of exported products turned it into a capital exporter. According to

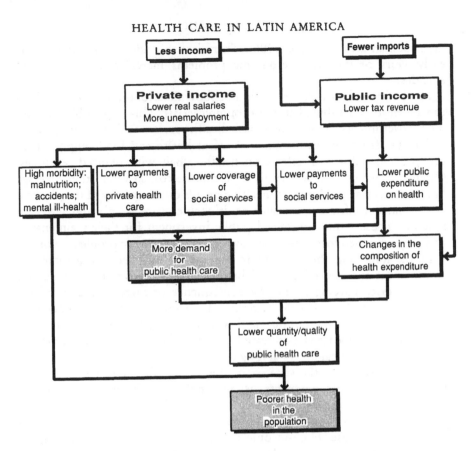

Figure 15.2 The relationship between income reductions, expenditure on health and public health levels
Source: Based on Musgrove (1987)

Inter-American Development Bank reports, interest on external debt, measured as a percentage of export values, increased. As a consequence, the majority of countries have been forced to reduce public expenditure and health care expenditure. Bolivia, Guatemala, the Dominican Republic and Surinam experienced drastic decrease. Only Haiti, Paraguay and Peru increased health care expenditure.

Figure 15.2 shows the direct effects of the crisis on health conditions. In the short term, the major effects will be the result of reductions in income and imports produced by the balance of payments adjustments. Economic recession will generally increase malnutrition. However, some effects are not yet evident in the records because of delays in collecting information and time-lags between health changes and their reflection in rates. Notwithstanding these factors, some diseases and age groups already show changes. For example, the São Paulo measles epidemic of 1984, following a delay of immunization, had great effects on the child population of the poorest metropolitan areas (Macedo 1986). There is also evidence that, during the early years of the crisis, economic problems increased the number of suicides.

Morbidity and mortality rates do not show deterioration in health which can be entirely attributed to the economic crisis. This is possibly because most governments support some basic health services such as immunization and focus efforts on programmes of infant health and food supply which can produce substantial effects on health, albeit with reduced resources. Often, vaccination programmes have been protected, so vaccination coverage has been increasing. Moreover, health depends on the quality of the physical environmental infrastructure such as drinkable water supply and sewerage systems. These require sustained public investment for their maintenance and they may therefore deteriorate over time as economic problems worsen.

In the long run, the decrease in investment will produce greater unemployment and lower incomes. The present unemployment levels reduce social security coverage, with the consequence of increasing demand for public health services which themselves have fewer resources. Reductions in family income have resulted in fewer payments for medical care. The immediate consequence is a decreased quantity of care received and generally poorer quality of health care among populations.

In many countries, the state is the major contributor to social security finances through its payments as an employer. Unemployment has a negative effect on social security finances, which depend on worker, employer and state contributions. Inequalities in access to health services are becoming worse. High-income people have access to a wide range of private health care systems. Many urban workers and some rural workers have partial access through jointly financed resources from the state, social security and direct payments to the private sector. However, the majority of the population, composed of informal workers, the unemployed, the marginally employed and the extremely poor, both urban and rural, have very limited or no access to the health care system. Their low income levels increase the risk of ill-health and death, and their basic necessities, including health care, should normally be provided by public systems which grow daily more inadequate in quality and quantity. The public health systems are almost invariably insufficient to meet their growing demand.

FOLK MEDICINE

Folk or traditional medical systems frequently coexist with modern systems. They are a product of centuries of experience and empiricism, magic and superstition. Folk medicine in Latin America is usually a synthesis of pre-Hispanic and colonial medicine produced as a consequence of the same conceptual, magical thoughts. The Spanish and Portuguese colonization imported medieval existential concepts, blended with magical components in the Catholic ceremonial, as well as in medical practices. Barbers went with the conquerors in the place of physicians. Religion reinforced the pre-Hispanic empiricism, and the Holy Office (the Inquisition) often limited the diffusion of new ideas which were the result of experimental sciences. So pre-Columbian cosmological vision survived as a result of cultural isolation, and the Cross, prayers and blessed water are components of the therapeutics prescribed by it.

Folk medicine involves a magico-religious concept in which diseases are viewed as supernatural phenomena produced by spiritual forces. These have an important role in folk societies because of the belief that humans live in a magic world, surrounded by mysterious forces producing diseases. Consequently, it is very difficult for folk medicine to provide a scientific explanation of diseases that satisfies the demands of academic medicine.

Perez de Nucci (1988) has identified a number of aetiologies in folk illness which include emanation or effluvium theory, loss of soul, magic and the breaking of taboos. Other authors classify folk aetiologies as conditions caused by mystics and by natural occurrences. The former include witchcraft, sorcery, evil wishes (conscious or unconscious) and the results of frights; the latter include effects of bad forces, falls, changes in temperature, and the like.

In this magic world, the most important person is the *curandero* or *curador* (the person who cures). *Curanderos* are physicians without a formal diploma and include spiritual or magico-religious healers, herbalists and technical specialists such as bone-setters. The *curandero* obtains knowledge by succession, by learning (from other *curanderos*) or by accident (for example, from a shock by a thunderbolt, commonly in desert latitudes). Perez de Nucci (1988) also includes predestination. Learning and exercise of the profession cannot by acquired solely by an individual's effort and intent. A spiritual element is required, because the *curandero* will have to see the future and make or destroy witches.

Another type is the 'charlatan' who appears to be a *curandero* but does not have the intention to cure, only to deceive. Charlatans use magic thought but do not believe in it. They use public shows with prominent people, and advertisements, and they are fraudsters, disguising their intentions. Unlike the *curandero*, the charlatan is always foreign, works in very expensive clinics and never works for free. Charlatans can grow rich on popular credulity. Stereotypes of traditional medicine include: holistic medicine, old and highly respectable people (except for charlatans) and folk illness. Traditional medicine can include traditional birth attendants (TBAs): *parteras* or *comadronas* (predominantly female).

The concepts and practices of folk medicine have stood the test of time and also academic critiques because they comprise not only a group of beliefs and superstition but also a structure of knowledge. The folk therapy can be herbal, mineral, animal, magical, empirical and intuitive. It has helped populations suffering from disease and war since pre-Hispanic times. Academic medicine would research the herbs and practices used in order to confirm their effects, but some traditional practices are valued because they belong to the cultural universe of folk belief. Folk medicine is a cultural phenomenon, not only a product of the current maldistribution of health care, personnel and resources.

Folk medicine exists in two essentially different environments. In rural areas, it exists where folk societies still survive. Here, the *curador* often has community approbation and is accepted by the authorities. In poor urban areas, the *chabolas* and *favelas* of the great cities, migrants have established and reconstructed part of their original culture. Here, folk medicine is protected from censure and repression by strategies through which it is clandestine or hidden from view.

However, in legal terms, Latin American health care is exclusive (monopolistic) to modern scientific medicine, or at best tolerant of folk medicine. Consequently, cultural confrontation frequently exists between scientific and folk medicine instead of complementary actions developing for popular health.

Health planning has considered neither the wishes of the people nor the 'magic relation' of health with environment. Consequently, it loses efficiency because patients are not understood. The magical universe of folk medicine must be considered in new models of health planning, and the *curador* has to be assigned an important role because (s)he participates in the magic thought and knows its mechanisms. Physicians in such settings have to appreciate the existence of folk thought, because health actions can generate a high resistance among people, and the objective of improved sanitation and environmental health cannot be obtained without popular co-operation. Technical actions without appropriate spiritual components will produce only partial results. This is an unresolved problem for Latin American societies.

CONCLUSION

The evolution of health in Latin America tends to differ from that in other poor areas of the world. The most important epidemiological issue in the region concerns the reasons for the changes in populations' health profiles. These changes are due not only to medical and technological progress, but also to the historical situation of countries and to the influence of world economic and political transformations, including economic crisis and structural adjustment. Some factors are therefore internal to specific countries and cultures, while others are regional, and the broad influence of world events also impact on health patterns.

The different stages of the epidemiological transition, which has occurred over more than a century in developed countries, often coexist in Latin

America, particularly in the middle-income countries (see Chapter 1). Nowadays, much morbidity and mortality from infectious diseases still persist. At the same time, chronic diseases, accidents and health problems relating to pollution and environmental degradation are increasing. As a consequence, the region has to develop different health care models in order to respond to different patterns of diseases. Infectious diseases need a community and a collective health care model, based on both environmental improvements, including sanitary engineering, and medical interventions such as vaccination. Cancers, heart diseases and accidents often need more and expensive technology for diagnosis and treatment, and individual as well as community health care.

Health depends on the interaction of numerous social, economic and political factors as well as on health care services. The external debt crisis and the economic adjustment policies within Latin America have changed the distribution of public expenditure and have transferred the responsibility for many social issues to the private sector. These changes in public expenditure and the privatization of health care and social security introduce new social relations between public and private health care organizations. Public health care has to compete with other sectors in order to obtain limited financial resources, and private organizations have fewer funds because of the reduced spending power and impoverishment of many groups. As a consequence, a great part of Latin America's population lives with inadequate health protection. Analysis of Latin American health care frequently shows only the deficiencies and risks, and forgets the explanation of the phenomena identified. The real issue in health and health care in the region is precisely the explanation of the causes and frequency of health problems, which without doubt include geographical factors.

REFERENCES

Macedo, R. (1986) *Brazilian Children and the Economic Crisis: Evidence from the State of São Paulo Revisited*, Rio de Janeiro: UNICEF.
Mesa Lago, C. (1988) *Desarrollo de la Seguridad Social en América Latina*, Estudios e informes de la CEPAL No. 43, Santiago de Chile.
Musgrove, P. (1987) 'The economic crisis and its impact on health care in Latin America and the Caribbean', *International Journal of Health Services* 17(3): 411–40.
Pan-American Health Organization (1982) *Health Conditions in the Americas, 1977–1980*, Scientific Publications No. 427, Washington, DC: PAHO.
—— (1988a) 'Resolución XV: desarrollo y fortalecimiento de los sistemas locales de salud en la transformación de los sistemas nacionales de salud', in *Informes finales: 100a y 101a Reuniones del Comité ejecutivo de la OPS: XXXIII Reunión del Consejo Directivo de la OPS: XL Reunión, Comité Regional de la OMS para las Américas*, Documento oficial 225, Washington, DC: PAHO, p. 60.
—— (1988b) *Los servicios de salud en las Américas: análisis de indicadores básicos*, Cuaderno técnico No. 14, Washington, DC: PAHO.

—— (1990a) *Health Conditions in the Americas, 1990 Edition*, Scientific Publications No. 524, Washington, DC: PAHO.

—— (1990b) 'Sistemas locales de salud (SILOS)', *Boletin de la Oficina Sanitoria Panamericana*, Washington, DC, pp. 109: 5–6.

Perez de Nucci, A.M. (1988) *La Medicina Tradicional del Noroeste Argentino: Historia y Presente*, Buenos Aires: Ediciones del Sol, Serie Antropológica.

Phillips, D.R. (1990) *Health and Health Care in the Third World*, London: Longman.

Verhasselt, Y. and Tuytschaever, L. (1987) *Mapping of Health-Related Indicators: A Possible Classification*, WHO/RPD/MAP/87, Geneva: WHO.

16

HEALTH CARE IN THE THIRD WORLD: AFRICA

Bose Iyun

INTRODUCTION

The evolution of Western health care in the Third World, including in Africa, followed the establishment of political administration *vis-à-vis* the introduction of modern economy in the colonized regions. To start with, the colonial administrators encountered severe health problems associated with the socio-cultural and physical environmental conditions of the African landscape, besides the roles of European and Arab immigrants in the introduction of some disease problems. For reasons probably connected with the lack of understanding and the need to take firm control of both the social and political affairs of Africans, the colonial administrators conceived the African traditional health care system as 'satanic', 'primitive' and 'unscientific' (Anyinam 1991). Its development was suppressed but not exterminated, and up to today this view is held by many African medical professionals.

In spite of the existence of Western health care for upwards of 100 years, most countries in Africa still show very severe health indicators. Besides the poor health profile exhibited by most African countries, inequalities in the distribution of health care resources and facilities are still pronounced. The scenario of the African health profile has in many ways been catastrophic since the introduction of the structural adjustment programmes (SAPs) forced on many countries by International Monetary Fund (IMF) and World Bank conditionalities.

This chapter intends to highlight some of the severe basic health indicators of African countries and the emergence of some worrisome retrogressive and transitional epidemiology *vis-à-vis* the problems associated with the lopsided provision and distribution of health care services. The chapter will also focus on attempts being made by various nations to reduce the continental health problems and to raise some of the issues that hinder adequate provision of health care in contemporary Africa. An attempt will also be made to indicate how to reduce some of these problems.

AN OVERVIEW OF THE HEALTH INDICATORS OF AFRICAN NATIONS

It is not difficult to see how precarious the health status of individuals is in Africa, in particular in sub-Saharan Africa, when we compare international health statistics. For instance, African countries make up 28 out of 38 countries that record very high under-5 mortality rates of over 140 per 1,000. Furthermore, *The State of the World's Children* reports for the early 1990s demonstrate that 34 out of 43 African countries still experience infant mortality rates of over 70 per 1,000 live births despite great improvements achieved since 1960.

Besides, Africa experiences very high maternal mortality rates, which again tell on its health care delivery system. In this, very few countries have made progress since 1960 and, indeed, maternal mortality rates appear to be worsening in a few countries, such as Malawi, Niger, Chad, Madagascar, Cameroon and even Egypt. Even though crude death rates are still high, most countries have experienced appreciable reduction, which has been reflected in the progress in life expectancy since 1960. On the other hand, the very high crude birth rates and total fertility levels maintained constitute great constraints on the general health of African women. The next three sections discuss trends in the development of health care on the continent from the pre-independence period to the structural adjustment encounter.

HEALTH CARE DURING THE PRE-INDEPENDENCE PERIOD

A preview of the development of Western health care delivery systems in most African countries reveals that health care was fashioned to deal principally with the health problems of European administrators, merchants, miners and others. Later on, African public servants and employees of various European organizations were included in the health care; this was probably due to the medical environment, which featured epidemics of highly infectious and killer diseases as well as the overwhelming public health problems (Schram 1971; Iyun 1978; Adeokun 1982; Anyinam 1991; Kalapula 1991).

Health care was therefore designed and distributed along racial lines. The establishment of the health facilities necessarily followed the distribution of political and economic positions or posts. Hence, inequalities in the distribution of hospitals in particular were inevitable. To this extent, locations that performed administrative functions, or that were economically productive, such as mining centres and transportation foci, attracted both health care facilities and resources.

As we would expect, the big urban centres received most of the health care resources. In many African countries, the disparities between urban and rural areas were pronounced. Intra-urban disparities were also apparent in the distribution of health facilities. However, the missionaries made concerted efforts to provide health care in the African rural areas and to reduce the inaccessibility of health services for rural dwellers (Schram 1971; Anyinam 1991; Kalapula 1991). Sometimes the medical environment has also helped in reducing inequalities in health care in some African countries. For instance, the occurrence of frequent epidemics in northern Nigeria provoked health actions in the 1930s that facilitated the establishment of health services in rural northern Nigeria compared with the southern part. In addition, political agitations by local communities often forced the colonial masters to establish health services in some rural localities (Iyun 1978).

By the time most African countries obtained their independence, four tiers of medical services, largely divided along socio-economic status lines, were in existence. These included: first point of contact (dispensaries, health posts, health centres) often run by local governments; secondary health care services (including general hospitals owned by central governments or by state governments in federations); and missions, private-ownership clinics and small hospitals run by individual private physicians, and profit-oriented and corporation- or company-owned facilities taking care of members of staff and some of their dependants. In parallel, traditional health care delivery systems organized across socio-cultural and religious lines were in operation in different African countries, and such systems are today often utilized simultaneously with Western health care.

HEALTH CARE AFTER INDEPENDENCE

By independence in the 1960s, in many African countries provision of health services had become a significant political issue. The need to eradicate some endemic and epidemic diseases such as sleeping sickness, leprosy, yaws, yellow fever and smallpox, in collaboration with international organizations such as the World Health Organization (WHO), the United Nations Children's Fund (UNICEF), the US Agency for International Development (USAID) and other non-governmental organizations (NGOs), had a prominent impact on the development of health care in most African countries. The promise of the provision of health care services and educational institutions was often at the top of the agenda of political parties during election campaigns.

Indeed, the actual locations of health facilities were often dictated by the whims and caprices of politicians. Thus, communities that could exert great political pressure often attracted health facilities. Eventually, the inequalities in the distribution of health care services created by the colonial administrators were further amplified by African politicians. Their role was critical in the apparent lack of direction and in the failure to implement the laudable programmes often stated in national plans. Emphasis continued to be laid on hospital services to boost coverage.

However, impressive strides were made in the training of more health workers of different categories. Indeed, between 1968 and 1978 many African countries were very close to the WHO recommendations on the ratio of health personnel to the population. Great efforts were also made to construct more hospitals, especially in the newly created administrative centres and other growing urban centres. Health intervention programmes of various kinds were embarked upon to control epidemics and improve on maternal and child health (Adeokun 1982; Anyinam 1991; Kalapula 1991). Some countries, such as Zambia, Sudan, Ghana and Benin, and some state governments in Nigeria, declared free health care (for the under-18s). Some countries, such as Zambia and northern Nigeria, experimented with the use of 'flying-doctor' services.

In spite of all these laudable efforts, the structure of the health care system in most African countries was very weak. Part of this weakness could be explained by faulty resource allocation among the different sectors of the economy and within the health sector itself. Most African countries gave less than 5 per cent of their annual budgetary allocation to the health sector. The situation is still much the same.

Resource allocation within the health sector was heavily skewed. More often than not, the lion's share of the budget was expended on staff salaries and wages (Ohiorhenuan et al, 1985). There was very little money left for the maintenance and running costs of facilities. Purchases of drugs and dressings also attracted an insignificant proportion and, to worsen the situation, wastage and irrational prescribing practices were common.

The precarious situation in the health sector was worsened by the unfavourable demographic and economic conditions on the continent. International comparisons such as *The State of the World's Children* illustrate the compounding problem of the health care delivery system. High population growth rates and an increasing proportion of Africans living in urban centres exert great pressure on existing health facilities, especially for maternal and child care services. Rapid rates of urbanization bring great pressure on hospital facilities in urban centres and an increase in the demand for hospital services. In many African countries, the private sector has reacted favourably to the increasing demand through the establishment of private clinics. This kind of process often brings about a shortfall in what public facilities can provide, since many public physicians opt out of government service.

By the end of 1980 it was obvious that dwindling government resources could not cope with the health demands of the inhabitants of the continent. The gross national product per capita of 62 per cent of the 43 African nations in 1990 was less than US$500. Meanwhile, inflation rates were over 5 per cent in 24 out of the 43 countries between 1980 and 1990. Events since 1983 have only helped to aggravate the precarious health care situation on the continent

HEALTH CARE SINCE STRUCTURAL ADJUSTMENT

By 1982 many African countries had started to experience severe difficulties in their debt repayment and the cost of debt servicing. Following the recommendations of the IMF and the World Bank, they started to reduce public expenditure, in particular on social welfare programmes including health and education. Unfortunately, the introduction of SAPs in many countries coincided with the implementation of primary health care (PHC), which aimed to foster adequate accessibility of health care to rural dwellers and the urban poor. This can also be regarded as the period when international organizations responsible for improvements in health care started to forge ahead with more rational and realistic programmes for the improvement of African health status (Djukanovic and Mach 1975), which was probably an opportunity for integrating traditional health care into the Western system at the primary level.

The intention here is not to repeat a list of the very adverse consequences of the implementation of SAPs on health care delivery in Africa, but to re-emphasize that the gains of the 1968–76 period have been lost in most African countries. Table 16.1 illustrates how many Africans still lack access to safe water and adequate health care services. While it is true that social and cultural influences affect the use of contraceptives, immunization and oral rehydration therapy (ORT), there is evidence that diminishing resources have been available to implement the most relevant components of PHC. In this connection, the lack of commitment on the part of many African leaders to implement PHC is probably a significant factor, in addition to the prevailing conditions of severe country indebtedness and aid dependency.

However, full implementation of the PHC strategy requires the mobilization of a lot of health resources to train and retrain health personnel, to improve the quality and quantity of health care at the points of contact and to integrate PHC firmly into the national health care delivery system. This calls for major expenditure. In this connection PHC goes beyond the provision of curative services. The strategy centres on improvement and maintenance of safe water supply, sewage and waste disposal. It involves the full immunization of children and mothers against immunizable diseases, the health education of mothers, under-5s nutritional improvements and drastic reductions in childhood killer diseases such as diarrhoea through ORT. Huge sums of money are required for implementation.

To worsen the situation in many African countries, besides the high costs of PHC implementation, the economic crisis brought about a 'brain drain' of health personnel, from highly skilled physicians (Iyun 1989; Loewenson 1990) to nurses and midwives – and it was the latter who were expected

Table 16.1 Access to basic health services in Africa, by country

Country	Population with access to safe water (%) 1988-9			Population with access to health services (%) 1988-90			Births attended by health personnel (%) 1989-90	Contraceptive prevalence (%) 1980-90	Year-old children immunized against DPT (%) 1989-90	ORT use rate (%) 1988-90
	Total	Urban	Rural	Total	Urban	Rural				
1 Mozambique	24	44	17	39	100	30	28	04	46	31
2 Mali	41	53	38	15	–	–	27	03	42	41
3 Angola	35	75	19	30	–	–	15	01	23	11
4 Sierra Leone	42	83	22	80	–	–	25	04	83	55
5 Malawi	53	97	50	80	–	–	45	07	81	14
6 Ethiopia	19	70	11	46	–	–	14	02	44	32
7 Guinea	32	55	24	47	100	40	25	–	17	63
8 Burkina Faso	69	44	72	59	51	48	30	01	37	16
9 Niger	61	100	52	41	99	30	47	01	13	54
10 Chad	–	–	–	30	–	–	24	01	20	10
11 Central African Rep.	12	13	11	45	–	–	66	–	82	20
12 Somalia	37	50	29	27	50	15	02	–	18	38
13 Mauritania	66	67	65	40	–	25	20	01	28	54
14 Rwanda	50	79	48	27	40	–	22	10	84	24
15 Burundi	38	100	34	61	–	–	19	07	86	30
16 Benin	54	66	46	18	–	–	45	09	67	45
17 Madagascar	22	62	10	56	–	–	62	–	46	11
18 Sudan	46	56	43	51	90	40	60	09	62	36
19 Tanzania	56	75	46	76	99	72	60	01	85	37
20 Namibia	–	–	–	–	–	–	–	–	52	–
21 Nigeria	–	54	54	66	85	62	40	06	57	35
22 Uganda	21	43	18	61	90	57	38	05	77	15
23 Gabon	68	90	50	90	–	–	80	–	78	10
24 Togo	71	100	61	61	–	–	15	12	61	33
25 Cameroon	44	50	39	41	44	39	10	02	56	22
26 Liberia	55	93	22	39	50	30	87	06	28	09

27 Ghana	57	93	39	60	92	45	55	13	57	21
28 Cote d'Ivoire	–	–	–	30	61	11	20	03	48	16
29 Zaire	33	59	17	26	40	17	–	01	31	40
30 Senegal	54	79	38	40	–	–	41	05	60	27
31 Lesotho	48	59	45	80	–	–	40	05	76	68
32 Zambia	60	76	43	75	100	50	39	01	79	87
33 Egypt	73	92	56	–	–	–	47	38	87	83
34 Morocco	61	100	25	70	100	50	29	36	81	14
35 Libya	94	100	80	–	–	–	76	–	84	60
36 Congo	38	92	02	83	97	70	–	–	79	13
37 Zimbabwe	66	31	80	71	100	62	60	43	73	77
38 Kenya	31	61	21	–	–	–	28	27	74	80
39 Algeria	71	85	55	88	100	80	15	07	89	26
40 South Africa	–	–	–	–	–	–	–	48	67	–
41 Tunisia	68	100	31	90	100	80	68	50	90	63
42 Mauritius	96	100	92	100	100	100	85	75	90	07
43 Botswana	54	84	46	89	100	85	78	33	86	66

Source: UNICEF (1990, 1992)

Note: Countries are ranked by UNICEF, in decreasing order of their under-5 mortality rate.

to implement the PHC strategy. Weekly changes in prices of essential drugs and dressings brought about by inflationary effects of the debt crisis further compounded the problem of providing acceptable, affordable and sustainable health care for the majority of the inhabitants of Africa.

For instance, Nigeria embarked on the implementation of the PHC strategy in 1986. Some fifty-two model local government areas (LGAs) were selected. A sum of N 500,000 Nigerian naira (equivalent to US$83,333) was to be made available to each LGA during the workshop period when each LGA drew up proposals. By the time the workshop participants returned to their localities, the Nigerian currency had been devalued. By the time the Federal Ministry of Health (FMOH) issued the first cheque for payment, the currency's value had been further reduced. As time went on, FMOH ceased to be in a position to help LGAs.

To reduce the costs of implementation of PHC, many African countries have experimented with the use of village volunteers or community-based delivery (CBD). Each country has experienced varied degrees of success in the use of village health workers (VHWs) or community health workers (CHWs) (Iyun 1989; Anyinam 1991; Kalapula 1991). Importantly, the prevailing economic situation does not augur well for free services from the village or market place-based volunteers. Hence the need to incorporate a 'human face' into the debt crisis, as always maintained by UNICEF, is imperative if the imminent collapse of health care delivery systems in most African countries is to be prevented.

THE WAY AHEAD

Health care is in a shambles in Africa. Access to recognized health care and personnel is diminishing, and this unfortunate situation has been capitalized upon by itinerant quacks who regularly visit rural dwellers, market traders, sea ports and airports to dispense dangerous and often date-expired drugs. Self-medication is on the increase, in particular for the treatment of some chronic diseases such as tuberculosis and diabetes.

Even though hard data are difficult to come by, the average expenditure of each household on health care is on the increase, because of overpriced imported and locally manufactured drugs and dressings. The continent's high inflationary rates cannot be accurately described, as the official figures are gross underestimates.

In the midst of all these problems that have almost strangulated Western health care in Africa, there is a resurgence of some infectious diseases such as yaws (in Nigeria in 1991), typhoid fever, yellow fever and cholera. Besides, a gradual but significant epidemiological transition is taking place, implying an increase in non-communicable diseases such as cardiac problems and cancers. The threat posed to the continent by AIDS is becoming overwhelming.

256

The most obvious way out of all these perilous situations is for African countries to strengthen their PHC, with highly formalized CBD to take care of not just children and mothers but also elderly people. The proportion of the elderly is increasing, and they are likely to face different social and health problems from the past when the African family was much more compact. The road ahead requires more than rhetorical pronouncements by African leaders and their condemnation by international communities in the West.

In an ever-shrinking world, with the knowledge that diseases do not require visas to cross international borders, and that their local impacts tend to have an effect on the wider society, lasting commitments to get Africa out of its precarious health situation must be explored. African countries must be discouraged from spending a disproportionate annual budget on the military and on arms purchases. Nor should African countries be made to spend every twenty-four hours in debt repayments and servicing enough money to put PHC on a proper footing. To build a solid foundation of PHC in the short run for both rural dwellers and the urban poor requires huge expenditure if it is to be effective and efficient. By implication, the inhabitants of Africa require PHC that will ensure safe water supply, immunization coverage of over 80 per cent of children and pregnant mothers, access to family planning and health education, and ORT for deadly childhood diseases such as diarrhoea. The international community should foster research into the impact of SAPs on health status in African countries, as such findings will lead to a more human perspective on the debt situation. The continent's disastrous decline in the health status of children, pregnant women and the elderly is an all-too-visible consequence of the debt crisis in Africa.

SUMMARY AND CONCLUSIONS

This chapter has focused on health care in African countries. It has emphasized the role of the colonial administration in disparities in the distribution of hospitals and lower-tier health facilities. Since independence in most African countries, the initial trend in provision of health care has been followed. Consequently, communities that can exert strong political pressure on African leaders have attracted more health care services.

Between 1968 and 1976 enormous progress was made in many African countries to construct more facilities, especially hospitals, but more importantly to train more health personnel of various cadres. The gains of this period have almost been lost in most African countries since 1982, after which many of the countries embarked upon structural adjustment.

The very adverse consequences of SAPs on the health status of Africans have taken the form of exorbitant prices of drugs and dressings, the 'brain drain' of trained health personnel, the resurgence of highly infectious diseases and the poor maintenance of basic social amenities such as water, sewerage and disposal facilities. There has therefore been a decline in the proportion of the continent's inhabitants with access to safe water and health services, while basic health indicators such as infant mortality are static.

The road ahead must involve the adoption and proper implementation of PHC. For PHC to be effective, huge sums of money and resources are required. It is apparent that African countries cannot put this in place while fulfilling international creditors' obligations in debt servicing. It is therefore imperative, to prevent the imminent consequences of further deterioration of health status in Africa, for international institutions and creditors to pay attention to how to promote health on the African continent.

REFERENCES

Adeokun, L.A. (1982) 'Nature, evolution and organization of health intervention programmes in Nigeria', *Quarterly Journal of Administration* 16(3–4): 151–67.
Anyinam C.A. (1991) 'Modern and traditional health care systems in Ghana', in R. Akhtar (ed.) *Health Care Patterns and Planning in Developing Countries*, New York: Greenwood Press, pp. 227–41.
Djukanovic, U. and Mach, E.P. (1975) 'Nigeria's use of two-way radio in the delivery of health services', in United Nations Children's Fund and World Health Organization, *Alternative Approaches to Meeting Basic Health Needs in Developing Countries*, Geneva: UNICEF/WHO.
Gish, O. (1977) *Guidelines for Health Planners: The Planning and Management of Health Services in Developing Countries*, London: Tri-Med Books.
Iyun, B.F. (1978) 'Spatial analysis of health care delivery in Ibadan city', PhD thesis, Department of Geography, University of Ghana, Legon.
—— (1989) 'An assessment of a rural health programme on child and maternity care: the Ogbomoso community health care programme, Oyo State, Nigeria', *Social Science and Medicine*, 29(8): 933–8.
Kalapula, E.S. (1991) 'Health care delivery patterns and planning in Zambia: colonial, populist and crisis interventionist approaches', in R. Akhtar (ed.) *Health Care Patterns and Planning in Developing Countries*, New York: Greenwood Press, pp. 199–225.
King, S.C. (1983) 'PHC: a response in support of a health revolution', *Development* 1: 29–30.
Loewenson, R. (1990) 'Health manpower issue in relation to equity in health services in Zimbabwe', *Journal of Social Development in Africa*, 5(1): 23–9.
Ohiorhenuan, J.M., Erinosho, O.A. and Iyun, B.F. (1985) 'Some financial aspects of the health care delivery system in Oyo State', paper prepared for the World Bank, Washington, DC, September.
Schram, R. (1971) *A History of the Nigerian Health Services*, Ibadan: University Press.
United Nations Children's Fund (1990) *The State of the World's Children 1990*, Oxford: Oxford University Press.
—— (1992) *The State of the World's Children 1992*, Oxford: Oxford University Press.
Walt, G. (1988) 'CHWs: are national programmes in crisis?', *Health Policy and Planning* 5(3[2]): 1–21.

17

HEALTH, ENVIRONMENT AND HEALTH CARE IN THE PEOPLE'S REPUBLIC OF CHINA

Fang Ru-Kang

The Chinese population accounts for one-fifth of humanity, so that a change in health, environment and health care in China can be said to greatly affect the world's health and development. This chapter summarizes a large amount of information to introduce the three aspects.

HEALTH

The health of Chinese people has improved along with the development of the economy. China historically was called 'the sick man of East Asia', but now China is one of the strongest countries in Asia, for example in terms of sports. Specific indicators of improvements in the health of China's people are as follows:

A decline in mortality

One of the indices for the improvement of health is a marked decline in mortality. Prior to 1949 China's population had experienced long-term high birth rates, high death rates and low natural growth. Although statistics are incomplete, it is estimated that birth rates were about 35 per 1,000 and the death rates about 25 per 1,000. Since 1949 the death rate has reduced sharply to 14 per 1,000 (1952), 10.04 (1963), 7.04 (1973) and 6.28 (1990) (Table 17.1)

Infant mortality is often accepted as an important indicator of a population's health. Before 1949 infant mortality was over 200 per 1,000 in China. Since 1949 it has rapidly declined: to 138.5 per 1,000 in 1954, 80.8 in 1958, 34.7 in 1981 and 19.6 in 1985. Infant mortality has shown a more rapid decline in urban than in rural areas; in Shanghai, for instance, it was 17.7 per 1,000 in 1952, 20.7 in 1962, 12.2 in 1982 and 11.3 in 1990.

Table 17.1 China: birth rate, death rate and natural increase (per 1,000)

Year	Whole country			City/urban			Country areas		
	Birth rate	Death rate	Increase rate	Birth rate	Death rate	Increase rate	Birth rate	Death rate	Increase rate
1949	36.00	20.00	16.00	–	–	–	–	–	–
1955	32.60	12.28	20.32	40.67	9.30	31.37	31.74	12.60	19.14
1960	20.86	25.43	-4.57	28.03	13.77	14.26	19.35	28.58	-9.23
1965	37.88	9.50	28.38	26.59	5.69	20.90	39.53	10.06	29.47
1971	30.65	7.32	23.33	21.30	5.35	15.95	31.86	7.57	24.29
1975	23.01	7.32	15.69	14.71	5.39	9.32	24.17	7.59	16.58
1980	18.21	6.34	11.87	14.17	5.48	8.69	18.82	6.47	12.35
1985	17.80	6.57	11.23	14.02	5.96	8.06	19.17	6.66	12.51
1989	20.83	6.50	14.33	15.98	5.74	10.24	22.28	6.69	15.59

Table 17.2 Leading causes of death in urban and rural areas of China, 1984

Rank	Urban			Rural		
	Cause of death	Mortality (per 100,000)	%	Cause of death	Mortality (per 100,000)	%
1	Heart diseases	124.64	22.65	Heart disease	158.53	24.57
2	Cerebro-vascular diseases	116.27	21.13	Cerebro-vascular diseases	98.78	15.31
3	Malignant neoplasms	116.18	21.11	Malignant neoplasms	97.02	15.03
4	Respiratory diseases	48.36	8.79	Respiratory diseases	78.50	12.16
5	Digestive diseases	23.76	4.32	Digestive diseases	36.39	5.64
6	Accidents	19.36	3.52	Poisoning	27.60	4.28
7	Pulmonary tuberculosis	10.17	1.85	Pulmonary tuberculosis	26.86	4.16
8	Poisoning	10.17	1.85	Accidents	20.80	3.22
9	Urinary diseases	9.48	1.72	Communicable diseases	15.32	2.37
10	Communicable diseases	8.14	1.48	Congenital diseases	10.40	1.62

Decreasing mortality from and incidence of illness

Epidemiological changes in the incidence of ill-health and death rates from illness in a population provide important indicators of health. Before 1949 endemic diseases and acute infections seriously threatened the whole country. After 1949 cholera, plague, smallpox, relapsing fever and venereal diseases were eliminated or drastically reduced. Other infectious diseases such as kala-azar, schistosomiasis, malaria, endemic goitre, Keshan disease and Kashin-Bek disease, which also seriously threatened people's health, were brought under control by means of prophylaxis and by effective treatment in large areas of the country. The cure rate of patients hospitalized for illness improved from 64.8 per cent in 1954 to 72.4 per cent in 1980 and 73.0 per cent in 1984. Table 17.2 indicates the relative causes of mortality in urban and rural areas in 1984.

Increase in average life expectancy

Prior to 1949 the average life expectancy of the Chinese people was only about 35 years. After 1949 the Chinese people's average life expectancy was extended to 57 years in 1957, 67.88 in 1982, 68.92 in 1985 and 69 in 1988. Average life expectancy in urban areas increased more rapidly. For example, in Shanghai it was 74.24 in 1985, 74.45 in 1987 (male 72.3, female 76.6) and 74.99 in 1990. This is near to the level of many developed countries. There has been an important concomitant ageing of the population which will provide a challenge to health and social services in the future (Kwong and Cai 1992).

In contrast to other countries, life expectancy of the people of China in the 1950s was very low, but by the year 1981 it was already well above that of most developing countries. It was higher than the average figures for Africa, Asia and Latin America and was approaching the average figures for Oceania (Table 17.3).

Increasing height and weight

Physical growth and development are important among the indices of the state of health in children and young people and are important foci for public health work. During the period 1978–80 a survey was conducted covering more than 200,000 children and young people aged 7–25 in 1,210 colleges and middle and primary schools in sixteen provinces and municipalities. Based on statistics available in eleven cities, the average height and weight of boys aged 7–18 increased between 1955 and 1979 by 5.6 cm and 3.25 kg respectively, and those of girls increased by 5.11 cm and 2.21 kg respectively. In other words, there was an average increase in height and weight of boys by 2.33 cm and 1.35 kg respectively every decade, and the average height and weight of girls increased by 2.13 cm and 0.92 kg respectively.

Table 17.3 Comparison of selected major indices: China and other regions

	Crude birth rate (1990).[a]	Crude death rate (1990).[a]	Under-5 mortality (1990)[b]	Life expectancy at birth (in years) Total	Female
China	21	7	42	70.1	71.8
High human development[c]	25	6	40	70.5	73.8
Medium human development (excl. China)[c]	25	7	71	66.2	68.1
Low human development[c]	39	12	152	56.8	57.6
All developing countries[c]	33	9	112	62.8	64.2

Source: based on data in UNDP (1992)
[a] Annual number per 1,000 of population.
[b] Per 1,000 live births.
[c] As defined by UNDP (1992).

Analysis of data obtained from various surveys shows a marked improvement in height, weight and chest circumference among children aged between 7 and 17 in 1979 as compared with 1975, with an average increase of 1.04 cm, 1.1 kg and 0.68 cm for boys, and 0.82 cm, 1 kg and 0.43 cm for girls. The biggest change was to be seen in children aged 13–16. Those who were examined in 1975 were born between 1959 and 1962 when the country was in a period of economic difficulty. The children were directly or indirectly affected by the unfavourable social and economic conditions, causing them to experience malnutrition and resultant poor development. Those who were examined in 1979 were born between 1963 and 1966 when the nation's economic situation and living standards had improved, apparently resulting in better physical development. This indicates that children's growth and development are closely associated with social and economic conditions, nutrition and levels of physical exercise. The most important factor appears to be nutrition, and the second is physical exercise. The effects of these two factors are particularly apparent during the accelerating growth phase of puberty. It follows that more attention should be paid to nutrition and physical exercise and training.

ENVIRONMENT AND HEALTH

Since 1979 environmental protection has become a basic policy for the government of China, and during the 1980s environmental protection became an issue of major importance for China's economic development. Environmental degradation has caused major losses, such as decreasing agricultural production, enforcing an increased awareness among decision-makers of the need for environmental protection. Although links between environment and health are not always clear, there is growing evidence that human-induced pollution and degradation are now major health hazards in China and many other countries.

In 1979 the Environmental Protection Law was passed and implemented. Environmental protection boards (EPBs) were set up in provinces, municipalities and counties all over the country. Additional legislation was also passed, including the Marine Environmental Protection Law, the Law on the Prevention and Control of Water Pollution, the Law on the Prevention and Control of Air Pollution and the Regulations on Prevention and Control of Ambient Noise Pollution. Furthermore, environmental quality standards for air and water have been set. The National Environmental Agency has the overall responsibility for the organization of environmental protection. At the national government level, an Environmental Protection Commission has been established, to co-ordinate environmental protection work between different ministries. The commission includes representatives from the ministries and works directly under the State Council of China.

Environmental impact assessments (EIAs) and an environmental levy system are used to control and limit pollution. Construction projects generating adverse effects on the environment must be preceded by an EIA, approved by the local EPB. All potential polluters' units are responsible for monitoring their discharges, and the results are reported to the local EPB. If the set standards are exceeded, the polluters' units are fined. The collected fees are used for the development of environmental protection techniques.

A grim situation

However, due to population growth, rising industrial production and the excessive use of natural resources, China still faces worsening environmental pollution. Qu Geping, director of the State Environment Protection Bureau, summed up the problem by saying that, while in some areas pollution is under control, the general situation is worsening. China is a vast country, and there are great regional differences due to variations in population density, degree of industrialization and climate, as well as natural conditions. A major problem is the shortage of drinking water, probably most urgent in China's northern provinces. In big cities the water table is dropping. One-fifth of the water reservoirs are contaminated due to uncontrolled industrial discharges and a lack of treatment plants for domestic wastes. Major environmental hazards to health include the following:

Air pollution

Air quality in many parts of China is poor. Coal is used to produce more than 70 per cent of China's total energy and is burnt in technically unrefined ways. Central heating in cities and gasification are far from widespread. The problem of air pollution is most serious in urban areas; some 15 million tonnes of sulphur dioxide were discharged in 1990. Moreover, some areas are affected by acid rain, such as the Chongqing–Yibin area in Sichuan, the Guiyang area in Guizhou, the Liuzhou area in Guangxi Zhuang Autonomous Region, Changsha in Hunan and Guangzhou in Guangdong. The affected areas extend from the south-west to the northern part of the country. Industrial pollution in some locations is also comparatively serious.

Water pollution

In 1990 a total of about 36 billion tonnes of water waste were discharged, including 21 billion tonnes of industrial wastes. Most waste water was discharged without treatment. Although the quality of the main water system is good, some sections are seriously polluted, implying a worsening trend. Polluted drinking water is also increasing. Due to the over-tapping of underground water, the water table has subsided in more than twenty cities. In addition, pollution has reduced the volume of potable water available.

Solid pollutants

The discharge of solid industrial pollutants and garbage in urban areas is increasing, and only a small part is being recycled. In 1990 solid industrial wastes amounted to about 580 million tonnes, only 26 per cent of which is reused. Industrial slag and garbage, which is piling up on city outskirts, is the second major source of solid pollutants.

Noise pollution

In general, the noise level in Chinese cities is comparatively high, with some statistics indicating that traffic noise on 70 per cent of city roads is as high as 70 decibels. Noise pollution is also serious in many residential areas; some 66 per cent of urban areas have ambient sound levels reaching 55 decibels. The noise pollution connected with industries and construction projects is on the rise.

A worsening ecology

Soil erosion affected 1.16 million sq km of China in the 1950s. Today, the figure has risen to 1.5 million sq km, and erosion affects about one-third of all cultivated areas. Each year, 100 million cubic metres more wood is logged before maturity. In addition, forest fires, plant diseases and pests and uncontrolled logging present serious problems. The quality of 51 million hectares of grassland, a quarter of the national total, has been adversely affected by desertification.

Environmental pollution is also prevalent in rural areas as a result of booming rural enterprises. Much of China's recent growth has been rural and in smaller towns, and even though in the past few years such areas have become one of the main forces of national economic development, those that lack proper leadership and planning generate a dangerous amount of pollutants.

Environmental protection strategies in China

Since the 1970s China has been aware of the harm brought about by environmental pollution and an unbalanced ecology. After a decade of research, a policy with Chinese characteristics has been devised to protect the environment:

1 China has established environmental protection as a basic national policy, setting guidelines for simultaneous planning and development of the economy, urban construction and environmental protection, co-ordinated with the nation's economic and social development plans.
2 China's environmental policy system has three major thrusts: 'prevent first and combine prevention with remedies'; 'polluters should pay the costs of clean-up'; and 'environmental oversight should be strengthened'.
3 China has established an environmental administration system with a unified leadership.

As a very populous developing country, China is currently unable to invest extensively in pollution prevention. The best it can do now is to strengthen environmental administration and to prevent pollution from getting worse.

Industrial contamination is the main cause of environmental pollution and ecological imbalance, and countering this is one of China's main goals of environmental protection. China has had some success in preventing industrial pollution. For example, industrial output value in 1988 had increased by 134 per cent over that of 1981, while discharge of industrial waste water increased by only 15 per cent. During the same period, waste water discharged as a result of industrial output by value had decreased by just over half, and heavy metals contained in waste water dropped by 64 per cent. Coal consumption had increased by 58 per cent, but the discharge of sulphur dioxide had risen only by 11 per cent, and that of smoke and dust had actually dropped by 1.2 per cent. Solid wastes from every 10,000 yuan (US$2,000) of industrial output value had dropped from over 7 tonnes to about 4.6 tonnes. Due to the effects of measures implemented, China has probably averted the worsening of its environmental pollution in the wake of robust economic growth.

However, such an achievement has not been easy. China has improved the protection of natural resources and the agricultural environment. In general, the quality of water in the main river systems and coastal areas is good. Major steps have been taken to prevent oil and heavy metal pollution, for example in the Bohai Gulf and Yellow Sea. Pesticide pollution has also been alleviated. Sixty per cent of alkaline land can now be used for agriculture, while 76 per cent of low-lying, flood-ravaged areas have seen improvement. Tree planting and afforestation have also progressed, especially in the shelter-belt areas of the plains, coastal areas and three northern areas. About 300 ecologically balanced agricultural centres have been established, along with 481 nature reserves covering 2.5 per cent of the nation's land surface. Since 1984 the state has established a list of 354 rare plant species and 527 rare animals threatened by extinction requiring regulatory protection. In the last few years China has also achieved initial success in its comprehensive programme to prevent pollution of the urban environment. In 1983 it was recommended that every province, city, department and factory should set its own environmental protection goals, which has since been implemented.

Chinese leaders have emphasized time and again the government's concern with global environmental problems. Their positive and serious attitude towards the world's environmental problem is affirmed by the fact that China has joined a dozen related international environmental protection treaties and conventions. China's drive to protect the environment has received widespread praise.

The Chinese government is confident that it will solve the nation's pollution problem within a reasonable time. While strengthening environmental protection though the full implementation of all environmental laws and regulations, the government is continuing to invest funds, allocating in 1979 about US$1.2 billion. The government expects to fulfil its 1992 target of stemming the rise in pollution, improving the environmental index in some major cities and areas, putting a halt to the worsening ecological balance and upgrading environmental protection. These should have major implications for China's public health.

HEALTH CARE

More than forty years have passed since the founding of New China, and the Chinese government has attached much importance to public health care. It has been regarded an important prerequisite for accelerating economic development. As the economy has improved, so too have central government, local governments and the communes increased their annual investment in public health care. This has resulted in great improvements.

In 1949, in a country so vast in territory and large in population, there were only some 3,670 medical and health institutions of varying sizes, with little over 80,000 beds. Medical and health workers numbered no more than 505,000. The few existing hospitals were located in major cities, and a

Table 17.4 Number of health institutions in China, 1949–89

	1949	1957	1965	1975	1985	1989
Total number of institutions	3,670	122,954	224,266	151,733	200,866	206,724
Hospitals	2,600	4,179	42,711	62,425	59,614	61,929
Sanatoria	30	835	887	297	640	651
Clinics	769	102,262	170,430	80,739	126,604	128,112
Specialized disease-control stations	11	626	822	683	1,566	1,747
Sanitation and anti-epidemic stations	0	1,626	2,499	2,912	3,410	3,591
Maternal and child care clinics	9	4,599	2,795	2,025	2,724	2,796
Medical research institutions	3	38	94	141	323	328
Total number of beds	84,625	461,802	1,033,305	1,764,329	2,487,086	2,867,000
Hospital beds	80,000	294,733	765,558	1,598,232	2,229,219	2,568,000

shortage of doctors and medical supplies left most rural areas without medical or health services.

Since 1949 the Chinese government has focused on developing health care in both urban and rural areas, with special emphasis on the latter. The number of medical and health institutions had grown to 206,724 by 1989, with more than 2.87 million beds. There were 4.21 million people employed, of whom 3.34 million were health care professionals. In the rural areas, almost every township now has a hospital, and 87 per cent of all villages have a clinic. A fairly complete health care network has taken shape, serving the needs of both cities and rural villages. The Chinese people have, for the first time in their history, been brought under effective protection against disease (Tables 17.4 and 17.5).

Table 17.5 Number of professional health personnel in China, 1949–89 (000s)

	1949	1957	1965	1975	1985	1989
Total	505.0	1,039.2	1,531.6	2,057.1	3,410.9	3,809.0
Doctors and assistant doctors of traditional Chinese medicine	276.0	337.0	321.4	228.6	336.2	370.0
Doctors of Western medicine	38.0	73.6	188.7	293.0	602.2	1,023.0
Medical auxiliaries	49.4	135.7	252.7	356.1	472.8	321.0
Nurses	32.8	128.2	234.5	379.5	637.0	922.0

China has not only made tremendous strides in increasing the number of health care facilities, but has also significantly improved their quality. As in other parts of the world, a fundamental shift in emphasis has occurred from treatment to prevention. The masses are being educated by professional health care personnel, and the incidence of infectious and endemic diseases has dropped sharply. Some diseases have been eradicated and others brought under control. Sanitation has improved. Great efforts have been made to popularize hygienic obstetric practices and to expand the scope of systematic health care for mothers and children. China has made great strides in integrating appropriate aspects of traditional Chinese medicine with modern techniques and pharmaceuticals. The following five themes in the health care system will be briefly discussed:

1 Public health organization.
2 Prevention and control of endemic diseases.
3 Prevention and control of parasitic diseases
4 Maternal and child care.
5 Health propaganda – the Patriotic Health and Sanitation Campaign

The Chinese public health system

Soon after 1949 the Chinese government laid down the policy of 'prevention first'. It rapidly established a fairly complete system of public health and epidemic disease control to fight epidemic diseases. Professional sanitation and anti-epidemic stations were set up according to administrative divisions at the provincial, autonomous-region and municipal levels (including provincial cities and countries), forming a three-tier system. Railways, communication facilities, factories and mines also set up their own sanitation and anti-epidemic stations according to their needs. These stations use theories of preventive medicine to control diseases, supervise sanitation work, spread knowledge about sanitation, train professional personnel and provide technical guidance for local disease-control programmes. They operate under the leadership of their respective administrative departments and professionally they are put under the leadership of sanitation and anti-epidemic stations at a higher level. Those in the industrial system are administered by their own administrative units and receive technical guidance from local sanitation and anti-epidemic stations. They operate under rules stipulated by the Ministry of Public Health, with equipment and facilities standards set by the state.

By the end of 1989 there were 3,591 sanitation and anti-epidemic stations in China, with a total staff of 171,800. Some areas have set up organizations for controlling occupational, parasitic and endemic diseases and for leprosy and tuberculosis, with a total of 1,747 institutions staffed by 478,700 professional personnel. A national preventive medicine institution was set up in 1983, and renamed the China Academy of Preventive Medicine in 1986. It carries out scientific research, trains personnel, provides technical guidance, supervises and monitors hygienic programmes, formulates hygiene regulations and standards and collects information about preventive medicine.

China has a vast rural area and a large rural population, for which the township epidemic disease control service plays a significant role. Ministry of Public Health provisions stipulate that one person should be in charge of disease control and health care for every 4,000 rural people. The main tasks of such personnel are: to help township governments to formulate plans for disease control; to spread knowledge about disease control and mobilize the people to carry out Patriotic Health and Sanitation Campaign activities; to control acute and chronic infectious diseases, parasitic and endemic diseases and occupational diseases; and to file epidemic reports and carry out immunization plans. They also provide guidance in and exercise supervision over environmental sanitation, food hygiene, industrial hygiene, school sanitation and radioactive hygiene; provide health care to women and children; provide technical guidance on disease prevention and health care to villages; and train village doctors.

Prevention and control of endemic diseases

Biochemical diseases such as Kashin-Bek, Keshan, endemic goitre and endemic fluorine poisoning are epidemic in parts of northern China, mainly in outlying hilly areas. In old China, these areas were impoverished, and poverty aggravated the effects of the diseases. After 1949 the Chinese government set up a leading group for the prevention and control of endemic diseases in northern China and sent medical teams to the endemic areas. In 1981 the leading group had its responsibility increased to include the whole country. Since then 28 provinces, autonomous regions and municipalities, 260 prefectures and cities and 1,500 counties have set up offices to take charge of the prevention and treatment of endemic diseases. At present China has 29 province-level endemic disease research institutes, with a total of 1,628 research workers. In addition, there is the National Endemics Science Committee, which has seven specialized groups.

At present goitre has basically been brought under control in fourteen provinces, autonomous regions and municipalities; Kashin-Bek disease is no longer spreading, and for many years there has been no outbreak of Keshan disease. In addition, the 10,000 water diversion projects completed so far have freed nearly 10 million people from fluorine poisoning. These represent significant advances in environmental threats to health.

Prevention and control of parasitic diseases

Here, the focus is on major parasitic or vectored diseases.

Schistosomiasis

Schistosomiasis is an important snail-borne disease spread in infected water bodies. According to surveys in the mid-1950s, schistosomiasis was found in Shanghai, Jiangsu, Zhejiang, Anhui, Fujian, Guangdong, Guangxi, Hunan, Hubei, Yunnan, Sichuan and Jiangxi. By the end of 1985 there were 371 endemic counties, cities and districts identified, with a total of 11.61 million cases and 14.3 billion square metres of snail-infested areas. Early research during the nationwide prevention and control programme found that 40 per cent of patients showed symptoms and about 5 per cent were in the late stage. Cattle and other domestic animals were also seriously suffering from the disease. The infestation rate was usually 20–30 per cent, with some areas at above 40 per cent. Schistosomiasis-endemic areas were divided into three topographical types: irrigation network areas, hilly areas, and lake and marshland areas. Snail-infested areas in irrigation network areas accounted for 8 per cent of the total area and for 33 per cent of patients. Lake and marshland areas accounted for 82 per cent of the area and for 41 per cent of the patients.

271

The elimination of snails is the key in schistosomiasis prevention and control measures. This is mainly done during water-related projects, including reclaiming wet areas, levelling the land, digging new ditches and filling old ones, changing rice paddies into dry crops, using more machinery to plough and press fields, building snail trenches and killing snails with chemicals. In lake and riverside areas, building dikes and using chemicals, section by section, are a method to kill snails. This has proved to be very effective along the banks of the Yangtze River in Jiangsu Province, where snails were eliminated from 18,666 hectares.

By the end of 1985 the disease was declared eradicated in Shanghai; and Guangdong, Fujian, Jiangsu and Guangxi had virtually eradicated the disease (over 98 per cent of snail-infested areas had been cleared of snails, and over 90 per cent of patients cured). Among the 371 endemic counties and cities, 110 have been declared clear of the disease, and in 161 it has been virtually eradicated. In the remaining 100 counties and cities, more than 50 per cent of townships and about two-thirds of villages have had it eradicated or virtually so. Of the 11.61 million schistosomiasis patients, 11 million have been cured. Eleven billion of the 14.3 billion square metres of affected areas have been cleared of snails, and 1.1 million of the 1.2 million head of cattle with the disease have been cured or destroyed.

Malaria

Malaria is one of the major parasitic diseases in China. In the 1950s there was an incidence of 5.07 per cent, or more than 30 million cases annually. Following efforts to eliminate the epidemic, control the environment and treat the disease by integrated methods, the infested area and the morbidity rate have continued to drop. In 1985 only about 473,000 cases were reported from January to November.

Since 1974 the five provinces on the Yellow and Huaihe river basins and the Yangtze–Hanjiang plain, the main tertian malaria areas (with morbidity accounting for about 80 per cent of the national total), have worked together to bring the disease under effective control. In 1985 the number of cases in these five provinces was 350,000, 97.4 per cent fewer than the almost 14 million reported in 1973, before the joint prevention and control measures were begun. The joint anti-malaria programme has now expanded to include fifteen provinces and autonomous regions, covering 390 million people or about 50 per cent of the population in malarial areas.

In August 1985 the Ministry of Public Health organized experts to make a spot-evaluation and assessment of the malaria control programme in twelve counties and cities, covering 20,000 square kilometres and containing a population of 3 million in Guizhou Province. Results showed that the disease was virtually eradicated in these places.

Filariasis

Filariasis is a disease found mainly in 864 counties and cities in 14 provinces, autonomous regions and municipalities; 54.1 per cent of counties and cities are endemic with the Bancroftian type and 26 per cent with the Malayan type. Some areas have both types. Following the prevention and control measures taken over recent years, filariasis areas have been drastically reduced and the severity of cases has significantly decreased. In 1983 Shandong Province, with a population of 75 million, was declared clear of the disease. By 1985, 660 counties and cities, accounting for 76.4 per cent of the epidemic areas, declared virtual eradication of the disease. The areas of Shandong, Guizhou, Guangxi and Shanghai have also virtually eradicated the disease.

Maternal and child care

Women and children make up about two-thirds of the total population of China, and their medical care has received considerable state attention. Special provisions and stipulations are made for the protection of women and children against disease in the country's constitution, marriage law and programmes for agricultural development and social development. This has ensured a steady improvement in maternal and child care organizations and health care.

General hospitals in all provinces, autonomous regions and municipalities have obstetrics and gynaecology departments, paediatrics departments and other appropriate health care sections. Some provinces and regions have specialized maternity and child care centres, obstetrics and gynaecology hospitals and paediatric hospitals. At the district and county levels, there are maternity and child care stations. The number of maternity and child care hospitals had increased from 80, with some 1,000 beds, in the post-liberation days to 272, with 24,443 beds, in 1985. The number of children's hospitals had increased from 5, with 139 beds, to 26, with 6,209 beds. The number of maternity and child care stations had risen from 9 to 2,724. These professional organizations, together with the primary medical care units in the urban and rural areas, form a vast network that provides adequate health care services to all women and children.

Professional workers in the fields of maternal and child care are mostly trained at medical universities and colleges. In addition, the Ministry of Public Health and the medical administrative departments in various provinces, autonomous regions and municipalities hold training courses and organize lectures or individual teaching to train maternal and child care workers in all related institutions. The number of maternal and child care workers is expanding, and their skills are improving. At present there are more than 60,000 professionals, of whom senior medical professionals account for about 25 per cent, secondary-level personnel for about 60 per cent and primary-level personnel for about 15 per cent.

Health propaganda: the Patriotic Health and Sanitation Campaign

The Patriotic Health and Sanitation Campaign is a health-for-all programme with distinct Chinese characteristics. It effectively carries out a 'prevention first' health policy. The campaign began in 1952 when Mao Zedong wrote: 'Get mobilized, pay attention to hygiene, reduce the incidence of diseases in order to raise the health standards.' Here, he laid down the methods, orientation, contents and purpose of the Patriotic Health and Sanitation Campaign. Since then the government has repeatedly issued instructions, circulars and other encouragements to further the steady development of the campaign. The country's constitution as revised in 1982 included the development of the campaign in its general principles, making it a basic national policy.

The campaign is organized and led by Patriotic Health and Sanitation Campaign committees at all levels, which are formed by leading members of government and mass organizations in the fields of urban and rural construction, planning, economics, agriculture and forestry, waterworks, education, sports, health, women's work and the Youth League. Leading members of the government take the posts of chair and vice-chair. This has provided the organizational framework for overseeing the campaign. Over the years the campaign has continuously disseminated medical and health knowledge and carried out health education, organized social investigations and research, and provided technical guidance in controlling vector insects and animals and improving urban and rural sanitation.

According to incomplete figures, there are now more than 200 popular publications on health, a dozen of which have a circulation of more than 100,000 and five of which have a circulation of between 500,000 and 1 million. There is also a health education handbook published nationwide by the government. In addition, there are no fewer than 400 separate kinds of booklets, leaflets and pictorials published in all parts of the country. The medical and health departments alone operate 30 popular science and telestrip studios, which made more than 80 films in 1985. At present, there are more than 90 health publicity and education centres in various provinces, autonomous regions and municipalities. Most prefectures and counties have set up health publicity sections or offices. It is estimated that there are now 12,000 people involved in health publicity work. A national health publicity network is taking shape which should gradually help to improve professional practices and public knowledge of ways towards good health.

Internationally, China actively participated in the activities of the International Drinking Water Supply and Sanitation Decade between 1981 and 1990. As the national action committee, the Central Patriotic Health and Sanitation Committee organizes all its activities in co-operation with other related departments.

CONCLUSION

China's progress in health care continues to improve the overall quality of life of the people. In many ways, it provides a model of change and enthusiasm for other Third World countries, working as it has done with vast numbers of people and very limited resources, and covering huge areas.

REFERENCES

Kwong, P. and Cai, G.X. (1992) 'Ageing in China: trends, problems and strategies', in D. Phillips (ed.) *Ageing in East and South-East Asia*, London: Edward Arnold, pp. 105–27.

Ministry of Health of China (1986) *Public Health in the People's Republic of China*, Beijing: People's Medical Publishing House.

State Statistics Bureau (1990) *Statistical Yearbook of China 1990*, Beijing: China's Statistics Press.

Tan Jianan *et al.* (1990) 'The progress in medical geography of China and its prospects', *Acta Geographica Sinica* 45(2).

Wang Shuze *et al.* (1986) *China Today: Health Care*, Beijing: China's Social Science Press.

Xu Dixin *et al.* (1988) *China Today: Population*, Beijing: China's Social Science Press.

Zhang Shen *et al.* (1990) 'Environmental pollution and its control in China', *Acta Geographica Sinica*, 45(2).

18

HEALTH AND DEVELOPMENT UNDER STATE SOCIALISM

The Hungarian experience

Eva Orosz

INTRODUCTION

The main purpose of this chapter is to contribute to revealing relationships between a population's health status, the health care system and socio-economic and political factors under state socialism. Both the state-socialist regime and the current transitional period in Hungary should be understood as specific types of socio-economic development. The chapter describes some basic tendencies; however, fundamental questions remain unanswered. It cannot provide a satisfactory explanation of concrete processes, through which deteriorating health status and characteristics of the state-socialist regime might be connected. The alarming deterioration in health status in recent years in many Eastern European countries may be interpreted as part of the failure of the state-socialist type of modernization, based on the fact that similar tendencies can be observed in a number of former state-socialist countries.

Hungary is a good example of a nation undergoing changes of historic significance: a transition to a multi-party democracy and a market economy. It is a painful process. Hungary and other countries in what was formerly known as Eastern Europe inherited not only a bankrupt economy but a great number of grave social problems. These include: deteriorating health status and decreasing life expectancy, decreasing living standards in many social strata, increasing poverty, excessive mental and physical strain on the population because of overwork, hazardous living and/or working conditions because of alarming rates of environmental pollution, increasing alcoholism and the highest suicide rate in Europe. These are but some of the problems facing Hungary. Furthermore, part of the cost of the economic transformation is likely to be an exacerbation of some of these problems and the creation of new ones such as mass unemployment. The contemporary Hungarian health care system must be understood in this context.

MAIN TRENDS IN HEALTH STATUS

Compared to advanced Western countries, both similar and different tendencies in health and demographic development can be discerned in Hungary over the last five decades. Similarities include decreases in the birth rate, ageing of the population and radical change in the causes of morbidity and mortality (a marked epidemiological transition). The total Hungarian population has been decreasing since 1980; a decreasing fertility rate and growing mortality rate have been decisive factors. Other Eastern European countries exhibit similar low natural increase; Bulgaria, the Czech and Slovak Republics, Poland and Romania all have rates of natural increase of 0.5 per cent per year or less, although Hungary's negative growth rate is an extreme even in the region.

Positive and negative trends have been simultaneously present in the health condition of the Hungarian population. Positive trends were obvious in the first two decades after 1945. The disappearance of famine, improvements in housing conditions and the extension of health care reinforced each other and resulted in a reduction of the grave problems of the inter-war period, particularly tuberculosis, epidemics caused by contagious diseases and high infant mortality. For example, in 1949 there were 9.61 deaths per 10,000 inhabitants due to tuberculosis ('morbus hungaricus') but there were only 2.51 in 1965. Infant mortality also decreased remarkably, from 156.9 per 1,000 in 1930 to 42.7 in 1961–5.

However, adverse trends have become dominant since the middle or late 1960s. In the mid-1960s the growth in male life expectancy stopped; it stagnated and slightly declined between the mid-1960s and mid-1970s, and it drastically declined after 1976–7. At present the life expectancy of males at birth is identical with the level of the late 1950s, and life expectancy at the age of 40 is no higher than it was in the late 1930s. These data can be interpreted as a 'social cost' of post-1945 Hungarian socio-economic development.

International comparisons of long-term trends in life expectancy show not only that *there are quantitative differences between the advanced Western countries and Hungary, but that their trends have diverged since the mid-1960s onwards.* This diverging trend is accompanied in Hungary by growing social and regional inequalities (Szalai 1986; Orosz 1990). From 1900 until the mid-1970s the trend in the average life expectancy of Hungarian men followed that in the present Organization for Economic Co-operation and Development (OECD) countries. While differences between OECD countries have further diminished in the past decade, Hungary since about 1975 has shifted to the periphery of the developed world in respect of the life expectancy of its males. This deterioration is brought out particularly boldly by comparisons with countries where life expectancy was lowest in the 1930s. Such countries had a specific

Table 18.1 Expectation of life at birth (ELB) and at age 40 for males

	1930	1950	1973	1983	1988
At birth:					
Finland	50.7	61.4	67.0	70.2	70.9
Japan	44.8	57.5	70.9	74.8	76.2
Portugal	–	56.1	65.2	69.1	70.1
Spain	48.4	59.8	70.3	71.8	73.6
Hungary	48.7	58.8	67.2	65.6	65.1
At age 40:					
Finland	27.6	28.9	31.8	–	29.1[a]
Japan	25.7	29.1	35.9	–	33.2[a]
Portugal	–	30.7	32.0	–	29.3[a]
Spain	–	30.4	34.5	–	31.7[a]
Hungary	29.1	31.1	29.5	–	24.8[a]

Sources: OECD (1987); WHO (1989, 1992); Demográfiai Évkönyv (Demographic Yearbook), Budapest
[a] At age 45.

semi-peripheral economic position, and their expectation of life at birth (ELB) was similar to that of Hungary (Table 18.1). Until the mid-1970s Hungary kept pace with life expectancy improvements in these countries. During the past two decades, however, a gap has opened between Hungary and the other countries surveyed.

The pattern of causes of mortality

The pattern of causes of mortality underwent radical changes between the late 1940s and the mid-1960s. In 1947, 11.8 per cent of mortality was caused by infectious diseases, 23.1 per cent by diseases of the circulatory system and 9.3 per cent by neoplasms. By 1964 the proportion of infectious disease had dropped to 3.3 per cent, whereas the share of the two other causes of death had been doubled: 50.9 per cent of deaths were caused by circulatory diseases, and 19 per cent by malignant neoplasms. Violent death had moved to third from sixth place. This pattern of mortality from the main diseases groups has only slightly modified since the 1960s. The proportion of mortality from infectious diseases has further decreased, and the largest proportionate increases have occurred in diseases of the digestive system and in violent deaths. More detailed analyses of mortality data show that some causes of death, such as the diseases of the coronary artery and the cerebro-vascular system, pulmonary cancer, suicide and cirrhosis, have been of outstanding significance in the worsening of male mortality. Cirrhosis has grown more than five times and pulmonary cancer has doubled as a cause of death since the mid-1960s. Social class differences for men are greatest for tuberculosis, diseases of the respiratory system and violent causes.

Differences between the mortality of males and females have also been increasing sharply during the past two decades.

Health inequalities

Inequalities in health are manifested quite differently in Hungary from the advanced Western European countries. In the latter, the health status of populations has improved, and generally inequalities have manifested in the fact that different strata of society have benefited unequally from these improvements. By contrast, in Hungary inequalities occur as the uneven distribution of the burdens of a process of deteriorating health status. The inequalities in health are most manifest between the more detailed occupational and educational groups. For example, the greatest improvement in infant mortality took place among mothers with higher education – consequently, the difference between the highly skilled and the other strata of society is greater now than it was in 1960. In 1960 infant mortality among mothers with education of fewer than eight classes was 1.9 times higher than among mothers with more than thirteen classes, and in 1990 it was 3.5 times higher. Infant mortality in the case of mothers with eight classes of education in 1960 was 48 per cent higher than among mothers with more than thirteen classes, and in 1990 it was as much as 84 per cent higher. The mortality rate of casual workers between 35 and 59 years of age was 9.3 times that of white-collar workers in public administration. This is an extreme case, but differences as great as two or three times often occur in this age group. The latest available mortality figures indicate that the above-described trends in mortality and life expectancy are likely to continue.

THE INFLUENCE OF SOCIO-ECONOMIC AND POLITICAL FACTORS

State-socialist economic development can be understood as an attempt to achieve modernization by the peripheral and semi-peripheral countries of Central and Eastern Europe. Arguably, the state-socialist system was doomed to fail because it was structurally unable to create the preconditions for further development in any sector of socio-economic life; it proved to be 'self-retarding modernization'. State-socialist economic development resulted in the exploitation and sacrifice of both natural and human resources including tragic degradation of the environment, deteriorating health status and underdevelopment of education and health care. The process feeds on itself: grave long-term social and environmental problems constitute serious obstacles to economic restructuring. The past five decades were, however, not homogeneous, and four periods may be distinguished.

In the first period, the state-socialist political and economic system evolved. Its main structural characteristics included the liquidation of political pluralism and the market economy; excessive penetration of the party-state into every sphere of society; strong centralization of power; a top-down institutional system; and, as part of the same process, the elimination of the institutions of civil society. The current system of health care administration also evolved according to these structural characteristics.

The development strategy was typified by excessive dominance of direct economic interests: in the 1950s the forced development of heavy industry devoured and diverted resources from infrastructure (including health care) and consumption by the population. The penetration of an important fallacy into political and partly into public thinking also contributed to the maintenance of this development strategy – the fallacy that health care and education belong to the non-productive sphere and thus were only consumers of national income.

The effects of socio-economic factors on health were contradictory. Some positive tendencies in health status in the 1950s have already been mentioned. However, long-term adverse effects of social processes on health status evolved and came to the fore some decades later. Large-scale social mobility provoked by forced industrialization and collectivization of agriculture, resulting in wide regional mobility (involving the inflow of rural population into urban areas), rapidly and radically changed ways of life. Uncertainty accompanying rapid changes, and the overtaxing of the adaptability of individuals, may be regarded as factors endangering the conditions of health.

The milestone of the second period was the reform of economic administration in 1968. After the late 1960s gradual erosion of the 'classical' system of state socialism took place in almost every sphere. A key feature of this process was the emergence and expansion of a 'second economy', leading or contributing to the formation of dual systems not only in the economy but also in other spheres including health care. Autonomy and entrepreneurial activities could – although in limited and distorted ways – widen in the everyday life of the people. Salai and Orosz (1992) suggest that:

> through the gradual expansion of informal production, people had started to build their lives on two pillars: one in the formal, and another in the informal sector. In this manner, a new way of life spread in Hungarian society, and two distinct clusters of motivation came to dominate people's daily activities.

These socio-economic processes, however, also had adverse effects on health. In the decade between 1960 and 1970 significant strata of society were able to improve their living standards (mainly their housing conditions) only by doing extra work. According to estimates, about 70–75 per cent of active earners pursued some extra work to supplement the income from their main job. Material improvement could be achieved by wide strata of society only at the cost of loss of health, distortions in lifestyles, sacrifice of their leisure and family break-up.

In the 1980s the multi-dimensional crisis of state socialism became visibly evident. Lasting economic crisis has had a complex, deleterious influence upon health. The burden of the crisis has been unevenly distributed in society. The growth of social differences has been manifest most obviously in the appearance of unemployment and in the growth of poverty.

In the 1980s policy-makers and society at large became aware of grave social problems. Environmental pollution increasingly threatened the health status of the population. Exposure to environmental hazards differs significantly between various regions of the country. It is an alarming fact that almost 40 per cent of the population lives in regions where air pollution is officially admitted (60 per cent of these people live in the Budapest agglomeration). During the past ten years the frequency of chronic bronchitic diseases has grown by 2.5 times, and the incidence of asthmatic diseases has multiplied by five. The number of settlements without clean drinking water has been rapidly growing. While 300–400 settlements had no healthy drinking water in 1970, this figure was about 700 in late 1980. Such growing environmental hazards are particularly alarming for the future health status of the young generation, as there can be as much as 20–40 years' delay between exposure to environmental risks and the development of disease.

Of the health-damaging elements of the way of life, researchers emphasize overwork, overstrained lifestyles, little active leisure, the consumption of alcohol and smoking. The per capita consumption of alcohol increased 2.2 times between 1950 and late 1980, and the proportion of people who died due to cirrhosis multiplied 7.3 times. According to estimates based on death caused by cirrhosis, there are between 400,000 and 500,000 alcoholics in Hungary. If family members are added, the lives of as many as 2 million people (one-fith of the population) are gravely affected by alcoholism. At present consumption, every sixth or seventh male and every fortieth or fiftieth female becomes an alcoholic during the course of his or her life. According to the 1984 micro-census, about 46–50 per cent of male manual workers and 37.6 per cent of white-collar workers smoke.

The cultural patterns that would promote tolerance and solution of conflicts are missing from society as a whole (in the life of individuals and from its institutions). The problems of alcoholism and suicide are partly related to this feature of Hungarian society.

There have been countries with high suicide rates in Central Europe since the beginning of the century. Hungary was already among the countries with the highest frequency of suicide during the inter-war period. International data show that the increasing frequency of suicide is not a necessary corollary of industrialization and urbanization. Most recently the rate of suicide has become higher in the villages, and suicide is more common in the disadvantaged strata of society.

The present period of transition to a market economy and multi-party democracy involves radical changes in political, economic and social structures which at present cannot yet be adequately grasped and predicted. However, some distressing tendencies may be noted. Transformation of Hungary's economy may be accompanied by huge and deepening social problems, including dramatically growing poverty and unemployment. In 1989 only a few hundred unemployed were registered, while in July 1992 more than 10 per cent of the economically active population were unemployed. These grave consequences of economic restructuring affect particularly those social groups whose health status has deteriorated during the past decades: the less educated, unskilled workers formerly employed in heavy industry, gypsies, and so on. Further increases in inequalities in health can be predicted.

THE HEALTH CARE SYSTEM AND SOCIO-ECONOMIC DEVELOPMENT UNDER STATE SOCIALISM

The 'residual' position of health care

There is a fundamental difference in the role of the state in Western welfare states and in state socialism. This is a pivotal point to the understanding of the differences of health care systems. Under state socialism, it was not welfare goals but direct production objectives that were excessively put into the focus of the redistributive activities of the state (Szelényi and Manchin 1987). As a consequence of the non-market economy, the state suffered from economic myopia. The support of loss-making companies enjoyed priority in the redistribution of incomes. Szalai and Orosz (1992) note that:

> Some 80 per cent of the yearly Gross Domestic Product (GDP) was concentrated in the hands of the state, and nearly 60 per cent of that huge sum has repeatedly flown through the productive sector of the economy – in the form of donations, subsidies and support to firms – to keep it alive. In this way it becomes understandable why the 'social budget' (the source of health services, culture, education, etc.) was repeatedly in a hopeless, impoverished and residual position.

A comparison of trends in health care expenditures in Hungary and in the advanced Western countries highlights fundamental differences. Between the early 1960s and the mid-1970s Western European countries experienced an expansionary phase: in the period of economic prosperity, health care expenditures increased 40–60 per cent faster than the gross domestic product (GDP). It made possible the modernization of the health care system, including the hospital infrastructure, and an extension of insurance coverage. The Hungarian health care system, however, has never experienced a similar (expansionary) period. During the 1960s and the 1970s the share of health expenditure in Hungary stagnated at 3–3.5 per cent of GDP – and the modernization of the health care infrastructure failed to be realized. Moreover, the utilization even of the available scant resources has been inefficient.

In Hungary, the share of health expenditure in GDP increased only in the period of economic recession, in the 1980s when GDP stagnated or declined. Health care expenditure amounted to 4 per cent of GDP in 1980 and to 5.3 per cent of GDP in 1990. In many cases it has not been enough even to maintain the previous quality of services. However, it has become evident that the grave structural problems of the health care system cannot be solved by solely injecting more money into unchanged structures.

In 1990 the social insurance fund took over health care financing from the state budget. The extreme instability of the financial situation of the social insurance fund is a factor at once forcing changes as well as narrowing the latitude for reforms. The main causes of instability in the social insurance fund are growing unemployment, an extensive and growing informal economy and the effort of the state as well as of private enterprises to increase the income of their employees under various pretexts which help them to evade the payment of taxes and contributions. Last but not least, a number of companies incurring losses simply do not pay contributions.

A grave dilemma of the current situation is that a reduction in the social burden of economic transformation would demand an increase in social expenditures, while from a macro-economic perspective the level of social expenditures as a whole as a percentage of GDP is too high compared to the less developed Western European countries. To enhance economic competitiveness would require the maintenance or reduction of social expenditure. However, to restructure the health care system would need an increase in resources devoted to health care. Public expenditure on health care must be understood within this context.

Main features of the state-socialist health care system

Central planning is not merely a financial technique for the distribution of resources, but it is a peculiar political relationship: a strong centralization of power, management on the basis of central order and an institutional system built from the top downwards. It involves a lack of autonomy among the different agents of the health care system, and a lack of institutions and regulated mechanisms for manifesting and reconciling interests of the main agents of the system. It is of fundamental importance to understand that the introduction of central planning did not simply mean organizational and financial change. Generally speaking, a health care system is an intricate web of institutions, relationships, mechanisms and attitudes, each with their roots in the historic development of a particular country. Central authorities, local governments, medical professionals, private entrepreneurs and different groups within society have different interests in, and attitudes to, the health care sector. Particular health care systems have been shaped and formed by competition, conflict, negotiations and consensus-building of these interests. In Hungary and in most state-socialist countries, these social and political processes and the institutional frameworks within which they operated, were eliminated and replaced by a system of central direction under a one-party state in the early 1950s. The local and county councils did not represent the autonomous actors in the health care sphere; their role was degraded to the implementation of central orders. Organizations representing the interests of the medical profession were liquidated. (The Medical Chamber was reorganized only in 1989.) Lay participation and control were entirely missing. The central authorities became the exclusive interpreters of needs and agents of priority-setting.

In contrast to the trend of health expenditures, the quantitative supply of physicians and hospital beds was considerably increased and became relatively larger than in most advanced Western countries. In 1989 the physician-to-population ratio was 29.4 per 10,000 people, and the number of hospital beds was 95.9 per 10,000 of population. By nature, central planning has an inherent bias to using quantitative increases in health care inputs as the sole measure of performance. As a result, under the pressure of chronic under-financing, an oversupply of physicians and short-term hospital beds has been accompanied by officially low medical-sector salaries, low levels of medical technology and a maldistribution of resources in terms of both specialism and geographical location.

Since the mid-1960s, as part of the previously described erosion of state-socialist structures, the central planning system in the health care sector had been diluted, and in the late 1980s it became increasingly disorganized. In practice, informal mechanisms of bargaining have determined the decisions between central authorities and the county councils, as well as between the health care administration and the physicians, rather than central orders. Personal relations have had a tremendous role in these processes of bargaining, as the institutions representing the interests of the main actors and regulated decision-making mechanisms have been missing.

The transformation of the political system in 1990–1 only produced the macro-level preconditions, but it was not sufficient to transform the decision-making processes within the health care sphere. Reform proposals continue to be worked out and decisions continue to be made without real involvement of the representatives of the main actors concerned, that is, solely by central bureaucracy. Decision-makers have not become aware that a necessary but not sufficient precondition for the success of any reform is to involve all the main actors: that is, to reform the process of reform itself. In this sense, Hungary and other Eastern European countries are captives of their former state-socialist political systems.

The dual system of Hungarian health care

The above description refers only to the official health care system. In fact, Hungarian health care has a dual system due to 'gratitude money'. The phenomenon plays a key role in understanding the Hungarian health care delivery system. In Hungary, health care is only officially and theoretically free. In reality, the majority of patients give so-called 'gratitude money' (in effect, a bribe) to general practitioners as well as to doctors in hospitals. The phenomenon can be interpreted as a tacit contract between politicians (via the health administration), doctors and society. Doctors accepted centralized bureaucratic management, and they also accepted a constant relative decrease of their official salary. However, 'gratitude money' is several times the amount of the official salary in the case of many doctors. For a long time, patients have been under the impression that adequate treatment is available only by payment of 'gratitude money'.

From a broader social aspect, it became characteristic of a number of fields of social life, including health care, that while there was no way of influencing the significant decisions within an organized, institutionalized framework, tacitly accepted and tolerated (although at first officially condemned) quests for individual strategies and advantages have operated since the mid-1960s. A 'shadow' health care system, similar to the 'shadow' or informal economy and functioning with different rules, has evolved alongside official health care. The relationships of suppliers of services and consumers are actually governed by this distorted market relationship. 'Gratitude money' creates an interest similar to that of fee-for-service remuneration. The doctors are interested in increasing numbers of patients and in a growing quantity of services, rather than in health promotion.

Consequently, it would be a misleading interpretation to present the current process simply as the transformation of an overcentralized state health care system into a 'public–private mix' system. It must be understood as the transformation of the current dual health care system – which is a symbiosis of an official state system and a 'shadow' private system – into a system in which both the 'public' and 'private' would be 'official', that is, in which 'shadow' health care would not exist. The question of *how* to transform the present dual system into a regulated 'public–private mix' is yet to be answered.

During the 1980s health policy became increasingly 'fiscalized' under the pressure of economic crisis, while the population's critical health status and the ailing health care system demanded a much broader perspective. A fiscal orientation, on the one hand, means that health care figured as an item burdening the budget from the point of view of governmental policy as a whole. On the other hand, issues of finance dominated almost exclusively the agenda of health policy itself. Health policy failed to address the basic structural problems of the health care system and the problems of health

285

status and inequalities of access to care. Emphasis was laid on the technique of financing, instead of on the structural problems hidden behind the inefficient use of resources. In spite of the change of political regime, several features and deficiencies of current health policy have their roots in that period.

The reorganization of the health care system is included in the programme of the new government which came into power with the 'systemic change' in 1990. There is an agreement on general goals: that the overhaul of the health care system must embrace its every component, including policy-making, ownership, financing, management, service structure, patients' rights and medical education. In reality, actual endeavours have narrowed to changing the methods of reimbursement of doctors and hospitals. Issues of vital importance, such as the effects of dramatically increasing unemployment and poverty on health, the long-term consequences of horrifying environmental pollution, and increasing social and territorial inequalities in health and access to health care, remain unaddressed by the government.

There is ironically a fundamental continuity in many attitudes. Under state socialism, social services, health care, education and environmental protection were considered 'non-productive' spheres and thus only as consumers of national income. This fallacy has prevailed, and the prominent role of these spheres in the reproduction of the labour force, in long-term economic growth and in competitiveness has still not been recognized by policy-makers.

The main characteristics of the present situation are an institutional vacuum, a lack of framework for consensus-making between the main actors and a lack of clarification of possible options for reform. Disorganization of the previous system has been accompanied by a struggle for the redistribution of power and new positions, implying a redefinition of values, and the development of new rhetoric. In this battle, short-term political interests and power relations play a more decisive role than considerations of the requirements of a workable health care system, let alone the improvement of the population's health status.

Ideological debates tend to formulate the question in an oversimplified way, namely: 'public' versus 'private'. The real issue, however, is what kind of 'public–private mix' should be developed under Hungary's particular circumstances. Privatization might be an important means to promote innovation, to revitalize rigid structures and to meet those special demands which public systems cannot and/or do not want to meet. The other side of the coin is that privatization might mean increasing inequalities and the institutionalization of a two-tier system. These phenomena – great inequalities and signs of a two-tier system – are not new; they existed under the previous state-socialist regime. The options, however, lie between containing or aggravating them. Many other countries in Eastern Europe face precisely the same dilemma.

CONCLUSION

This chapter has considered the relationships between the population's health, the health care system and socio-economic and political factors, in the phases of setting up, erosion of and crisis in state socialism. International comparisons of long-term trends in life expectancy show not only that there are quantitative differences between the advanced Western countries and Hungary and other similar countries, but also that their trends have diverged from the mid-1960s onwards.

With regard to health policy, in theory the main goals of health systems are to provide physical and social environments that promote and maintain health and provide access to health care services which prevent and cure diseases. In reality, these objectives have not represented the main driving force in the operation of the Hungarian system at any period: the means have been replaced by the ends.

Ongoing reorganizing of the health care system must be understood as an element of the realignment of the whole social welfare sector. The historical transformation of the economy means a grave challenge for the social welfare sector. This is in part because economic recession is considerably greater, and the social costs of economic transformation are far higher, than expected; and in part because of the collapse of the economic basis upon which the earlier social welfare system was built.

A key question in the ongoing socio-economic transition is what kind of relationship would be appropriate between the economy and social welfare sectors (including health care) under Hungary's specific circumstances. Ideologically, the 'social market economy' is the catch-phrase; however, the meaning of the 'social market economy' is unclear and much disputed. At least three different approaches can be discerned: (1) The 'social market economy' might mean the notion of welfare state implying universal rights, entitlement to wide-ranging and decent benefits, dominance of public financing and control in the welfare sector (including health care). (2) On the contrary, it might mean an endeavour to create a free market with the minimum possible public intervention, when social assistance (such as public financing in health care) should be targeted only to the poor. (3) The concept of a 'social market economy' might be used primarily to legitimize a hierarchical welfare system with a strong position for the state bureaucracy. It is yet to be seen which meaning of the 'social market economy' will become dominant in Hungary or in other Eastern European countries as they deal with their emergence from state socialism.

REFERENCES

Organization for Economic Co-operation and Development (1987) *Financing and Delivering Health Care*, Paris: OECD.

Orosz, E. (1990) 'Regional inequalities in the Hungarian health care system', *Geoforum* 2: 245–59.

Szalai, J. (1986) 'Inequalities in access to health care in Hungary', *Social Science and Medicine* 2: 135–40.

Szalai, J. and Orosz, E. (1992) 'Social policy in Hungary', in B. Deacon (ed.) *The New Eastern Europe: Social Policy Past, Present and Future*, London: Sage.

Szelényi, I. and Manchin, R. (1987) 'Social policy under state socialism: market redistribution and social inequalities in Eastern European socialist societies', in M. Rein, G. Esping-Anderson and L. Rainwater (eds) *Stagnation and Renewal in Social Policy*, Armonk, NY:

World Health Organization (1989, 1992)*World Health Statistics Annual*, Geneva: WHO.

19

SOUTHERN EUROPE: HEALTH IN AREAS THAT HAVE UNDERGONE RECENT DEVELOPMENT

Demographic and epidemiological issues

Cosimo Palagiano

INTRODUCTION

From a geographical point of view Southern Europe can be considered as the three major peninsulas that are washed by the Mediterranean Sea. In addition to these, the Mediterranean region includes Mediterranean France and European Turkey, which do not belong to the Mediterranean peninsulas, and Portugal, though it is not washed by the Mediterranean Sea. The area of the peninsulas of Southern Europe is about 1.5 million sq km, with a population of 153 million inhabitants, giving a population density of 107 inhabitants per sq km. The most important countries are Portugal, Spain, Italy, the former Yugoslavia, Albania, Bulgaria and Greece. However, these countries are quite dissimilar socially, economically and politically. While sharing the geographical characteristics noted below, some now have to contend with very serious problems of identity and development.

The countries of Southern Europe that have undergone considerable modern recent development are Portugal, Spain, Italy and Greece, and they at different times became members of the European Community. These countries have in common certain geographical characteristics: mountainous morphology; high insularity; a lack of pit coal and iron ore; and their development was generally retarded compared with that of other Western Europe countries, in both specific economic sectors and social facilities. They are territorially unbalanced internally. The major economic parameters of the countries are listed in Table 19.1. This chapter focuses on the epidemiological profiles and health care systems of these four major countries, for which data are available.

Gross national product (GNP) is very different in the four major countries of Southern Europe: for example, Italy's highest GNP is over

Table 19.1 GNP per capita and average annual rate of inflation, Southern Europe

Countries	GNP per capita US$ (1989)	Average annual growth 1965–89 %	Average annual rate of inflation 1965–80 %	1980–9 %
Portugal	4,250	3.0	11.7	19.1
Spain	9,330	2.4	12.3	9.4
Italy	15,120	3.0	11.4	10.3
Greece	5,350	2.9	10.3	18.2

Sources: based on data from the International Monetary Fund, UN and World Bank

three times that of Portugal's. The average annual growth in GNP between 1965 and 1989 is about 3 per cent in three of the four countries. Spain had the lowest average annual growth from 1965 to 1989. The average annual rate of inflation between 1965 and 1980 is 11.4 per cent for all countries, but the average for 1980–9 is higher, because of the high rates in Portugal and Greece.

In each of the four countries considered in this chapter there are certain regions which show better economic conditions than others; for example, the Lisbon district contributes 25–30 per cent of the national wealth of Portugal. In Spain, the richest regions are Cataluña and the Balaeric Islands and the poorest is Extremadura, while in the middle economically there are the Basque Country, Valenciana, Murcia and Galicia. In Italy, the richest regions are those in the north-west, but in recent times other regions have developed their economies, such as the regions of so-called 'Middle Italy', which includes the central and north-eastern regions of the country. Greece is the most underdeveloped of the current EC countries, with a large trade gap, despite its tourism revenue. In Greece, the richest regions are the urban areas of Athens and Salonika. Over the past twenty years the richest and poorest regions of these countries have converged, but the reduced differentials in economic levels are due more to the crises of the overdeveloped regions than to the improvement of the underdeveloped ones.

DEMOGRAPHIC ISSUES

The distribution of the total population by age groups in the four countries is relatively uniform, with the exception of Italy (Table 19.2). In Italy the young groups show lower percentages, while the elderly groups of 50 to 64 show the highest values. The Greek population shows a comparable trend to that of Italy, without considering the under-15 population, while Portugal shows the opposite, with a larger presence of young population. The demographic situation in any country depends on the annual rate of increase, the birth and death rates, infant mortality rate, life expectancy and

Table 19.2 Area, resident population and age structure, Southern Europe

Countries	Area (sq km)	Resident population (millions)	Resident population (per sq km)	Age groups (%) <15	Age groups (%) 65+
Portugal	91,985	10.305	112	23.0	12.1
Spain	505,954	38.969	77	22.0	12.4
Italy	301,263	56.557	192	17.3	14.1
Greece	131,957	9.740	74	21.6	13.3

Source: UN data

Table 19.3 Selected demographic data, Southern Europe

Countries	Annual rate of population increase (%) (around 1990)	Birth rate[a]	Death rate[a]	Infant mortality rate[b]	Life expectancy at birth (in years)
Portugal	0.7	12.0	9.3	14.2	75
Spain	0.4	11.2	7.9	10.5	77
Italy	0.1	9.7	9.1	8.8	76
Greece	0.4	10.6	9.5	12.2	77

Source: UN and UNICEF data
[a] Annual number per 1,000 of population.
[b] Per 1,000 live births.

migration. Information on the major demographic characteristics is given in Table 19.3

The range in infant mortality is very high (Figure 19.1), with big differences between Italy and Portugal. Greece also shows a high infant mortality rate. Birth rates are more uniform, with the exception of Italy's lower rate. However, Italy's lower rate is an average for the north and south of the country, the former having a lower rate and the south a much higher value. Mortality rates are more homogeneous. Spain has the lowest crude mortality rate of the four countries. Italy's population is showing quite remarkable ageing and now has one of the lowest rates of natural increase in Europe. This in itself will have important implications for health services provision and social care in coming years.

THE EPIDEMIOLOGICAL PICTURE IN THE FOUR COUNTRIES

The four countries have similar epidemiological profiles; all have passed through the earlier stages of epidemiological transition, and today the main causes of death are neoplasms and cardio-vascular diseases, while infectious diseases are strongly diminishing. However, there are some differences in detail, which should be emphasized, in relation to the incidence and prevalence of the various causes of death. We consider here eight main

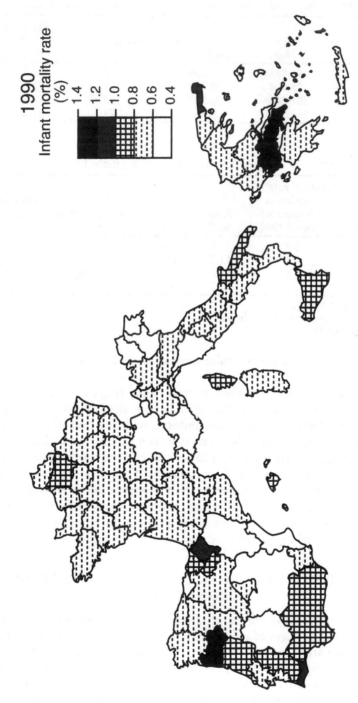

Figure 19.1 Southern Europe: infant mortality rates, 1990 (per 1,000 live births)
Source: based on European Community data

causes of death for the following three age groups: under 20, 20–59, and 60 and over. The main causes of death being considered are infectious diseases, neoplasms, endocrine and metabolic diseases, diseases of the blood, diseases of the nervous system, cardio-vascular diseases, diseases of the respiratory system and diseases of the digestive system.

Portugal

In Portugal the cardio-vascular diseases were the major cause of death in 1988 for the aged, while many people aged 20–59 years fell victims to neoplasms. Other major causes of death are, in order, the respiratory, the digestive and the endocrine diseases. If we consider the trend in the causes of death for all age groups during 1983–8, it may be noted that mortality from cardio-vascular diseases is increasing, but with some decreasing fluctuations during the years 1986 and 1987 and with a rate of 420 to 440 deaths per 100,000 people. The second major cause of death, the neoplasms, had a rate of 170 deaths per 100,000 people which is steadily increasing. While mortality for respiratory diseases varies from year to year, that for digestive diseases decreases and that for endocrine diseases increases. For these three diseases, the mortality rate is 69, 49 and 25 per 100,000 people respectively.

A geographical outline of this epidemiological situation can be given, taking into account the economic condition of the country. As noted above, Portugal has started to become an industrial country, but it has had to confront many socio-economic problems related to the backwardness of its economic framework. The infectious diseases still present a moderate percentage of mortality; mortality from cardio-vascular diseases is much higher than mortality from neoplasms; mortality from cardio-vascular diseases does not tend to decrease but, rather, to increase; mortality from neoplasms tends to increase, but it is still much lower than mortality from cardio-vascular diseases. This epidemiological framework can be explained with reference to Portugal's socio-economic situation, which is not yet fully industrialized. In fact, in many countries which are or were in a pre-industrialized situation, we can note a similar epidemiological situation: a strong, steadily increasing prevalence of mortality from cardio-vascular diseases and a lower, but constantly increasing, prevalence of mortality from neoplasms. In addition to this, countries such as Portugal show a prevalence of endocrine, respiratory and digestive diseases, which can lead to death. There is a suspicion of a bad quality of both life and nutrition which might underlie the high rates of these disorders. The speed of epidemiological change is important, as it has been occurring over only a few decades. This challenges the health care system and public health, which have to cope with emerging chronic conditions as well as a continuance of infectious ailments.

With regard to health care facilities, in 1988 there were 229 hospitals (147 general and 82 specialized), 371 health centres, 2,450 pharmacies and 261 treatment posts in all. In the Lisbon region there are 65 hospitals, and in the Porto region there are 32 hospitals. The number of physicians per 100,000 inhabitants varies from 228.0 in 1983 to 261.1 in 1988, with an increase of 14.5 per cent . The number of beds per 1,000 inhabitants varies from 5.1 in 1983 to 4.7 in 1988, showing a slight decrease. Health facilities need to be improved in the immediate future. This situation can also in part explain the high rates of perinatal, neo-natal and infant mortality, which were respectively 15.2, 8.6 and 13.1 per 1,000 in 1988. This is well above what would be accepted in the north-west countries of Europe, and more on a par with some Eastern European countries.

Spain

The major cause of death in Spain is cardio-vascular diseases, which have a rate of 330 to 360 per 100,000 people. The second major cause of death is the neoplasms, which have rapidly increased during 1981–6 from a rate of 150 to 180 per 100,000. The third major cause is the respiratory diseases, which have a rate of around 64 per 100,000. Other frequent causes of death are the digestive diseases, which are constant around a rate of 40 per 100,000, and the endocrine diseases, which have increased slightly from 20 to 25 deaths per 100,000.

In terms of health facilities, in 1987 in Spain there were 886 hospitals with 171,155 beds (4.4 beds per 1,000 people). 48.7 per cent of hospitals are general, with 68 per cent of beds; other hospitals are the *hospitales especiales de corta estancia* surgical hospitals (of which there are 142, with 9,731 beds), children's hospitals (6 hospitals, with 852 beds) and maternity hospitals (6 hospitals, with 1,415 beds). In Spain in 1987 there were 4.39 hospital beds per 1,000 people, but, as in many aspects, there are strong regional variations in their distribution. The peripheral regions, such as Andalucía, Extremadura, Galacia and La Mancha, tend to have below-average numbers of beds per 1,000 people.

With regard to medical staff, in 1989 there were 143,803 physicians, 9,433 dentists, 35,141 chemists, 11,595 veterinaries and 157,194 hospital attendants. The physician-to-population ratio is on average 3.69 per 1,000. The most favoured regions are Aragon (4.84), Comunidad de Madrid (4.79) and Navarra (4.70), while the least favoured are Ceuta y Melilla (2.61), Castilla La Mancha and Extremadura (2.72) and Galicia (2.90). It is important to note that most physicians reside in the major cities: 76 per cent of the physicians of Aragon work in Zaragoza; 62 per cent of the physicians of the Comunidad Valenciana work in Valencia; 58 per cent of the physicians of the Basque Country work in Vizcaya; 28 per cent of the physicians of Andalucía work in Seville. There are thus important urban–rural differentials in availability of health care, which inevitably affect the quality of care and services that people receive.

Italy

A major feature in Italy is how neoplasms have become the most important cause of death for relatively young people. The nervous ailments and respiratory diseases were decreasing as causes of death in the mid-1980s, while the cardio-vascular diseases show the contrary trend. The third main group of causes of death are the infectious, endocrine and digestive diseases, but the infectious and digestive diseases are decreasing and the endocrine and blood diseases are increasing. In the 25–59 age group, neoplasms are now the main cause of death, followed by cardio-vascular diseases, although these are decreasing. The digestive diseases also have importance as causes of death, but they too are decreasing. For the over-60 age group, the relationships between the neoplasms and the cardio-vascular diseases are different from in the previous age groups. Cardio-vascular diseases are the main cause of death for people over 60, and the neoplasms are the second major cause of death. However, neoplasms are increasing and cardio-vascular diseases are decreasing. A large percentage of the deaths from cardio-vascular diseases are probably due to poor diagnosis. In fact, the cardio-vascular diseases are frequently the final cause of death of a sick person.

The major increase as a cause of death and also as a disease among the neoplasms is lung cancer. The incidence is 27,000 cases in men and 5,000 in women. This pathology is, of course, strongly linked to cigarette consumption, which was increasing to ten years ago in the male population and is still increasing in the female population. This is therefore a disease strongly influenced by social habits and one which health education, taxation on cigarettes and control on tobacco advertising may help to reduce.

Italy has marked regional variations in the distribution of health care facilities, in many ways reminiscent of those in most developing countries. The health facility data in Table 19.4 indicate the extreme differences among the different regions. Expenditure diminishes from north to south, while physicians and attendants are insufficient in most Italian regions. The ratio of beds to inhabitants is generous above the number considered excellent, 6 beds per 1,000 inhabitants. But three southern regions have fewer than 6 beds per 1,000 inhabitants.

The difference in both quantity and quality of the health facilities in Italy underlies a 'sanitary migration' from the southern to northern regions and also abroad, particularly to France. An 'attraction rate' index for both middle-quality and best health facilities is very high in the northern regions and very low in the southern regions, as shown in Table 19.5. The attraction rate is the ratio between the percentage of emigrants (the ratio between the number of the residents in the region being admitted in another region and the hospitalized people resident in that region) and the percentage of immigrants (the ratio between non-residents and the total number of hospitalized people in the region).

Table 19.4 Italy: health facilities by region

Regions	Health expenditure per inhabitant	Physicians	Attendants (per 1,000 people)	Beds
Piemonte	1,179	3.77	3.98	6.87
Valle d'Aosta	1,215	3.36	4.33	5.23
Lombardia	1,114	4.22	4.10	7.67
Trentino Alto Adige	1,193	3.47	4.74	8.25
Veneto	1,214	3.91	5.00	9.34
Friuli Venezia Giulia	1,257	4.22	6.29	11.60
Liguria	1,382	5.85	5.26	9.53
Emilia Romagna	1,360	5.10	5.26	8.54
Toscana	1,283	5.08	5.81	8.10
Umbria	1,223	5.21	4.61	5.98
Marche	1,333	4.52	4.98	8.67
Lazio	1,207	6.25	3.37	8.01
Abruzzo	1,171	5.32	4.63	9.30
Molise	1,137	4.55	3.97	4.01
Campania	1,092	4.85	2.90	5.52
Puglia	1,081	4.08	2.95	8.16
Basilicata	953	3.58	3.18	6.73
Calabria	987	5.03	3.58	6.11
Sicilia	1,112	5.03	2.67	4.91
Sardegna	1,060	4.65	3.83	6.78
ITALY	1,170	4.72	4.03	7.50

Source: Italian government data

This sequence generally ranks the regions of Italy from north to south, because it is well known that the northern regions are provided with the best health facilities. In addition, the central regions are provided with good health facilities because of their inclusion in the 'attraction area' of the capital, Rome. The sanitary migration from the Italian regions which are not provided with good health facilities focuses not only on well-provided Italian regions but also on France, where in some hospitals a large proportion of patients are Italian. This migration for health care provides an interesting and important commentary on the spatial unevenness of provision in Italy, and one which policy-makers should address.

Greece

The main cause of death in Greece is cardio-vascular diseases, with 48,273 deaths, and a crude mortality rate of 496 deaths per 100,000 people. The second major cause of death is the neoplasms, with a mortality rate of 192 deaths per 100,000 people. Thirty-one per cent of neoplasms are cancer of the digestive organs and peritoneum, and 25 per cent are malignant neoplasms of lung and chest. It is interesting to note that neoplasms are now more relevant than the cardio-vascular diseases for the ages 20 to 59. This

Table 19.5 The 'attraction rate' of the Italian regions by rank

Regions	Medium health assistance[a]	Best health assistance[a]	
Liguria	19.19	Liguria	16.25
Lombardia	17.62	Veneto	9.81
Emilia Romagna	6.71	Lombardia	8.63
Lazio	6.51	Friuli Venezia Giulia	5.15
Friuli Venezia Giulia	5.85	Emilia Romagna	4.36
Veneto	1.52	Lazio	4.14
Toscana	1.46	Marche	1.56
Marche	1.20	Abruzzo	1.26
Umbria	0.78	Toscana	1.08
Valle d'Aosta	0.56	Umbria	0.49
Trentino Alto Adige	0.45	Piemonte	0.38
Abruzzo	0.35	Puglia	0.16
Piemonte	0.31	Basilicata	0.13
Puglia	0.24	Campania	0.09
Molise	0.18	Molise	0.08
Campania	0.13	Valle d'Aosta	0.07
Basilicata	0.02	Trentino Alto Adige	0.05
Sicilia	0.02	Sicilia	0.02
Sardegna	0.01	Sardegna	0.008
Calabria	0.003	Calabria	0.005

Source: Crotti (1991)
[a] The difference between 'medium' and 'best' health assistance (*specialità a media* and *a elevata assistenza*) is explained by an Act of the Ministry of Health of 13 September 1988.

appears a strong progression in mainly industrialized countries. The aged are the most affected by the other important causes of death in Greece, such as those of the respiratory system, a rate of 51 deaths per 100,000 people, those of the nervous system, the endocrine and metabolic systems and the blood and blood-forming organs. The infectious and parasitic diseases have only a limited importance as causes of death in Greece, but there are still to be found some infectious diseases, indeed ones that have disappeared in the developed countries. As in some middle-income countries, infectious ailments in Greece still contribute an important, if declining, cause of mortality.

In terms of health-care facilities, in 1987 in Greece there were 454 hospitals, with 51,745 beds, 15,205 physicians, 1,815 midwives, 14,608 trained nurses and 11,031 practical nurses; 223 (49.12 per cent) hospitals are general, 193 (42.51 per cent) are specialized, and 38 (8.37 per cent) are combined. As regards the legal form of the hospitals, 136 are legal public entities, including hospitals under the authority of the Ministry of Health, Welfare and Social Insurance and the Ministry of Interior, and hospitals belonging to the Social Insurance Organization, the Maternity and Child Welfare Institution, the Army Share Fund and the universities. Private-sector facilities comprise 59 per cent of the total number of hospitals and have 30.73 per cent of the total number of beds.

In 1987 there were in Greece 5.31 hospital beds per 1,000 people (with 8.41 beds per 1,000 people in Greater Athens) and 3.42 physicians per 1,000 people (6.13 physicians per 1,000 people in Greater Athens). The capital obviously has a considerable concentration of medical facilities and personnel, in many ways reminiscent of the situation in some developing countries. The ratios of beds to inhabitants and physicians to inhabitants of Macedonia show 5.33 and 3.21 respectively, approximately the average values of the country. However, over 70 per cent of the physicians of Macedonia work in the city of Thessaloniki, again a typical urban bias of a middle-income country.

CONCLUSION

The countries of Southern Europe show epidemiological patterns that for many reasons are similar but with an important range of characteristics. The common characteristics include the decline of mortality from infectious and parasitic diseases; high mortality from cardio-vascular diseases and neoplasms; and the weakness and the uneven distribution of many health facilities, due also to the conflicting interests of the private and public hospitals and clinics. However, there are differences among the four countries considered. In the less developed countries of the group, Portugal and Greece, considerable mortality is still due to many infectious, respiratory, endocrine and digestive diseases. Mortality through cardio-vascular diseases is greater than through neoplasms in the less developed countries, while in the more developed countries of the group, Italy and Spain, cancers are now a more important cause of death. Health facilities are improving in all four countries but they are still insufficient in the less developed of them. Important issues of availability and accessibility, especially regionally uneven distribution of health care facilities and urban–rural differentials, remain to be addressed and present policy-makers with a major challenge for equity and equality.

REFERENCES

Cohen, A. (1991) 'La dynamique géographique de la mortalité en Espagne', *Espace, Populations, Sociétés* 1: 135–41.
Cónim, C. and Carrilho, M.J. (1989) *'Situação Demográfica e Perspectivas de Evolução: Portugal, 1960–2000*, Cadernos 16, Lisbon: Instituto de Estudos para o Desenvolvimento.
Crotti, N. (1991) 'La migrazione sanitaria: valutazione ed inter pretazione del fenomeno: problemi organizzativi ed assistenziali', in M. Geddes (ed.) *La salute degli Italiani, Rapporto 1991*, Rome: NIS.
Decroly, J.-M. and Vanlaer, J. (1991) *Atlas de la Population Européenne*, Brussels: Edit. Univ. Bruxelles.
Faus-Pujol, M.C. (1991) 'Morbidité–mortalité en Espagne', *Espace, Populations, Sociétés* 1: 127–34.

Lizza, G. (1991) *Integrazione e Regionalizzazione nella Cee*, Milan: Angeli.
Meneghel, G. (1991) 'Mortality and socio-economic conditions in Italy', *Espace, Populations, Sociétés* 1: 173–80.
Palagiano, C. (1989) *La Geografia della Salute in Italia*, Milan: Franco Angeli.

20

HEALTH AND DEVELOPMENT: RETROSPECT AND PROSPECT

David R. Phillips and Yola Verhasselt

HEALTHY DEVELOPMENT – AN ELUSIVE GOAL?

The purpose of this concluding chapter is to review some of the major issues that have been highlighted by the various authors in the book. A major underlying theme, identified in Chapter 1 and pursued in many subsequent chapters, is that there is not a simple, one-way correlation between 'greater development' and 'better health'. While to some extent this is a matter of the definition of the two concepts, both largely unquantifiable and each with a subjective element, it is nevertheless increasingly clear that development improvements often equate with *different* health and not simply *better* health. In the majority of countries with higher and increasing real incomes per capita, there is often (but not invariably) a longer average expectation of life, but this longer life may have attendant health problems. It can also have differing social care problems related to family and community care, the availability of pensions and social support and appropriate accommodation. These are, it is true, emerging as issues in many middle-income countries, but there is often a degree of family support that continues to encompass older people in less developed economies.

This is only one of many crucial issues touched on in the book and to which we believe increasing attention will need to be given by policy-makers and care-providers in the future. As editors of the book, and as researchers in the field of health and development, we have identified at least ten key issues which have emerged over the past decade or so and which will assume increasing importance in the current decade and the early years of the twenty-first century. These are not identified in order of priority; they are for the most part not single, self-contained phenomena, but they are in many ways interlinked 'syndromes', to employ a medical phrase. They include a broad 'environmental' syndrome, an 'economic–debt–structure' syndrome, an 'ageing population structure' syndrome, a 'development policies' and an 'infectious disease' syndrome, to name a few. We turn our attention in the closing pages of the book to these and to the underlying characteristic of many of them which relates to poverty of countries, of individuals and of concern. Many of the issues have inherently geographical and environmental components, and geographers involved in health research, and working in a wide range of types of institutions, have been at the forefront of researching many of them. The geography of health has become an important sub-discipline working at the integrative boundary of many social and biomedical sciences (Verhasselt 1993). We hope that the value of this book will lie in part in the compilation for ready access of a good range of examples of such work.

ENVIRONMENTAL CHANGE AND HEALTH

Environmental change and health is perhaps one of the major media issues in developed countries with respect to individual health and the future ecological systems in which population groups will grow up. Chapter 2 provides an important overview of the impacts of global environmental change on health. Focus has been on direct impacts which this tends to have on fair-skinned people living in warm-climate countries or in countries in which elements of solar radiation are likely to become more hazardous as protective atmospheric layers are eroded by industrial emissions. Much concern has revolved around the consequences of ozone depletion and increasing UV radiation for the incidence of skin cancers and eye disorders, particularly cataracts. Higher average temperatures associated with global warming may lead to sea-level rises, the loss of food-producing areas and threats from flooding in low-lying areas, particularly in places such as Bangladesh. Global warming may also considerably extend the areas in which certain vectors of infectious diseases will be able to survive.

Many other chapters concern the local effects of environmental change and micro-environmental degradation. In urban areas, all residents are often being exposed to higher levels of potentially toxic substances, and this appears to increase as income grows (a surrogate measure for development). Up to a point, environmental problems appear to increase but they then often decrease as knowledge of them grows and the ability and willingness to devote resources to them also increases.

Indeed, the 1992 *World Development Report* of the World Bank has a particular focus on development and the environment. Figure 20.1 summarizes some of the ideas that may impinge on health from the World Bank report. It shows that there is probably not an inevitable and invariable relationship between income levels and particular environmental problems. It does, however, suggest that things will often get worse, environmentally speaking, before they improve, as development progresses. The report emphasizes that policies can be chosen or designed to harness the positive links between development and the environment and they can also be targeted remedially, to ameliorate specific environmental problems. Damage to the environment has at least three, often interrelated potential costs to

302

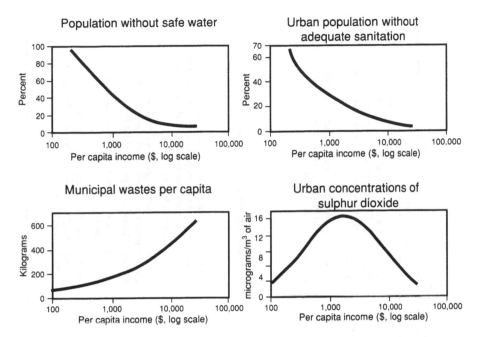

Figure 20.1 Environmental indicators at different country income levels: problems
may worsen or improve with income growth; some worsen, then improve
Source: Based on World Bank (1992: 11)
Note: Estimates are based on cross-country regression analysis of data from the 1980s

present and future human well-being. Human health may be harmed, economic productivity may be reduced and 'amenity' value lost. The World Bank (1992) identifies a range of health and productivity consequences of environmental mismanagement (Table 20.1). It is evident that environmental problems vary across countries and with the nature and stage of industrialization, so each country needs to assess carefully its own priorities.

One of the difficulties today of identifying precise links between such environmental problems as localized pollution, emissions and background radiation is that individuals are exposed to conditions for very differing degrees and at different times of their lives, and their individual and group genetic characteristics differ and affect their bodies' responses to environmental threats. The residential and occupational mobility of many people today also makes it very difficult to determine exactly which environmental risk factors they have been exposed to and for what time period. In addition, personal lifestyles, eating and drinking habits, smoking or substance abuse will all be factors affecting overall dangers to health. Perhaps the only certain thing is that we do not currently have a great enough understanding of the relationships between environment and health, at the global, local or individual scale. Our knowledge is constantly increasing, but the picture

303

Table 20.1 Principal health and productivity consequences of environmental mismanagement

Environmental problem	Effect on health	Effect on productivity
Water pollution and water scarcity	More than 2 million deaths; billions of illnesses a year attributable to pollution; poor household hygiene and added health risks caused by water scarcity	Declining fisheries; rural household time and municipal costs of providing safe water; aquifer depletion leading to irreversible compaction; constraint on economic activity because of water shortages
Air pollution	Many acute and chronic health impacts: excessive urban particulate matter levels are responsible for 300,000–700,000 premature deaths annually and for half of childhood chronic coughing; 400 million–700 million people, mainly women and children in poor rural areas, affected by smoky indoor air	Restrictions on vehicle and industrial activity during critical episodes; effect of acid rain on forests and water bodies
Solid and hazardous wastes	Diseases spread by rotting garbage and blocked drains; risks from hazardous wastes typically local but often acute	Pollution of groundwater resources
Soil degradation	Reduced nutrition for poor farmers on depleted soils; greater susceptibility to drought	Field productivity losses in range of 0.5–1.5% of GNP common on tropical soils; offsite siltation of reservoirs, river transport channels and other hydrologic investments
Deforestation	Localized flooding, leading to death and disease	Loss of sustainable logging potential and of erosion prevention, watershed stability and carbon sequestration provided by forests
Loss of biodiversity	Potential loss of new drugs	Reduction of ecosystem adaptability and loss of genetic resources
Atmospheric changes	Possible shifts in vector-borne diseases; risks from climatic natural disasters; diseases attributable to ozone depletion (perhaps 300,000 additional cases of skin cancer a year worldwide; 1.7 million cases of cataracts)	Sea-rise damage to coastal investments; regional changes in agricultural productivity; disruption of marine food chain

Source: World Bank (1992:4)

becomes, ironically, more complex rather than clearer, with the exception of certain specific ailments.

HEALTH IMPACTS OF STRUCTURAL ADJUSTMENT AND THE IMPACT OF ECONOMIC CRISIS

Chapters 1 and 3 of the book outlined some of the health issues involved in 'structural adjustment' policies which for the past decade or so have been increasingly imposed by the international community (of rich nations) on to indebted Third World countries or countries which looked as if they might run into economic deficits. These adjustment policies have often been euphemisms for the retrenchment of public spending on social and welfare programmes, the most important ones of which have generally been health. However, education (arguably more important to long-term health than is the provision of curative medical services) has also generally suffered in these austerity programmes. Their impacts have been wider, however, and have diminished individual and families' incomes, and their ability to buy food, essential clothing and shelter, and to obtain health and welfare care. Children have dropped out of education to try to work at pitifully young ages, and both families and governments in many poor and middle-income countries have been unable to redress this. Crucially, reduced public expenditure has often cut back valuable environmental health, sanitation, infrastructure and housing programmes. Inhabitants have been left in deteriorating local living conditions in which many health gains of earlier years are being lost.

As public and private incomes and expenditure have fallen, so, too, has public and private provision of health care retracted, among many other things. Chapter 13, for the poorer Third World, and Chapter 15, for Latin America, point to the pervasive relationships between public expenditure cuts, falling public health provision and reduced personal incomes, tied inexorably with reduced expenditure on essentials. However, such classical recessionary cycles are by no means confined to the Third World. For many years, the economics of centre and right political groups in Europe and North America have stressed the virtues of countries and families 'living within their means'. At a national level, this has resulted in reductions (or insufficient expansion) of many forms of public provision in Britain and other countries in Europe and Australasia, with New Zealand providing a salutary example of cut-backs in welfare expenditure in the 1990s recession. In the United States, President Clinton pledged in 1993 to rein back on many forms of expenditure on health, welfare and education, at the same time increasing taxes to cut financial deficits. Both the rich and poor in such societies feel the impact of these policies, although, of course, the poor most directly exhibit their health effects. While wealthier nations may impose structural adjustment on the Third World, the international financial markets can punish developed countries which do not play by fairly rigid set of non-inflationary expenditure rules.

Cuts in public expenditure on health and welfare, and in public subsidies for services, medicines and food, are part and parcel of this. However, particularly since the mid-1980s, there have been attempts to ration health services or to recoup expenditure of them (depending on one's point of view). So-called 'cost recovery' schemes have been tried in a number of developing countries and particularly in Africa following the 1987 Bamako Initiative. Many fear that these policies will exclude the poorest and most needy from essential drugs and medicines. However, it seems that forms of cost recovery are here to stay in many systems. Indeed, the wealthier the country, with some notable exceptions mainly in the Organization of Petroleum Exporting Countries, the more likely it seems that some element of user charge will be levied even for 'free' essential health care. Alternatively, individual responsibility, often via private insurance schemes, is increasingly praised by governments. Evidence shows that the most needy in health terms are the most frequently ineligible: elderly people, the chronically ill, the poor. In the Third World, as discussed in Chapters 1 and 3, adjustment 'with a human face' has been advocated, but its success has been limited.

DEMOGRAPHIC AGEING AND HEALTH TRANSITION

Many chapters of the book have concerned themselves directly or indirectly with the health effects of populations ageing and placing increased, or at least different, demands on health care services. Many authorities in fact see the health transition, or what has more specifically been called the epidemiological transition, as the most important aspect of population ageing. Whether it is a cause or an effect of population ageing (healthy populations live longer, etc) is of research interest to many demographers. However, the practical effects of epidemiological change are that there is an increased amount of chronic and/or degenerative disease in most populations. This generally may need some direct medical attention, but, more often than not, social care and community support are needed. This requires a changed orientation in many health care systems towards primary care and support and a redirection of funds and of medical training away from high-tech hospitals and centralized facilities which may only treat, expensively, a small proportion of acute illness. Some of the adverse consequences of ageing can be reduced by prevention of diseases and disability – a point that should be stressed in policies, actions and training related to public health and welfare services.

Many Third World countries are currently experiencing rapid demographic change. In most countries, birth rates and total fertility rates are reducing; and in many (but not all), child survival rates are steadily improving. Slowly but surely, more peole are living to older ages, and a group of countries in East and South-east Asia and Latin America now have substantial proportions of their populations aged over 60. Indeed, in more than twenty-five developing countries in 1992, average life expectancy is over 70 years. Average life expectancy in developing countries is now 63 years, 17 years more than in 1960.

In some countries, this population ageing has occurred remarkably rapidly. For example, in a period of under fifty years, Japan will by about 2020 have gone from having a young population structure to one of the oldest in the world. The rapidity of this ageing, in a country of very high labour and infrastructure costs, has led to great uncertainty and almost fear of the future. Much research in Europe and North America has tended to focus on the demographic causes of population ageing and their economic consequences (see, for example, United Nations 1992). However, there are a growing number of researchers interested in social gerontology who are pointing out the health care and social care consequences as well as the need to recognize the changes in families' and communities' abilities to play major caring roles (Ory and Bond 1989; Warnes 1989; Philips 1992).

Many Third World countries, and many of their cities, do still face the twin threats from infectious diseases such as cholera, malaria and hepatitis, as well as AIDS (discussed below), while, at the same time, they are meeting an ever-rising tide of chronic conditions: heart diseases, cancers and cerebro-vascular disorders. They are having to deal with these twin problems with generally depleted resources and in the face of ever-increasing medical costs. Often, medical curricula are hopelessly inadequate for the social care aspects of epidemiological change. Personnel are still fighting many of yesterday's infectious foes, yet to be defeated, but rarely are they addressing today's emerging health problems. In many richer developed countries, complacency about infectious diseases has led almost to an inability to tackle some such as HIV in which changes in lifestyle and behaviour are, to a large extent, likely to be the only foreseeable means of halting spread. The faith of residents in many developed countries that high-tech medical intervention can cure all ills has been sorely tested by recent developments, including the emergence of highly drug-resistant strains of tuberculosis and certain other infections.

It is likely that addressing causes of morbidity (illness, disability, handi-cap) will assume a proportionately increasing importance as compared with mortality in the furture. Many causes of death are sequels to occasionally long periods of illness. These have huge personal and societal consequences, but, because of the fight to defeat killing infectious diseases and in part because of difficulties in measuring morbidity, this has in the past received proportionately less attention. Like mortality rates, morbidity shows con-siderable social and spatial variations, to do with genetic differences and exposure to health hazards because of where people live and the work they

307

do. Murray and Chen (1992) explain that both self-reported morbidity and observed (assessed) morbidity will become increasing foci of attention. Greater numbers of people, especially in the developed world, are reporting themselves as ill or disabled. The explanations for this are complex and may include greater awareness of disease, earlier diagnosis, access to better health services, lower population mortality rates and, almost certainly, welfare payments and raised public expectations of bodily health. These combined phenomena have huge implications for the future of health and health care under changing conditions of development.

VULNERABLE GROUPS AND THE POVERTY COMPLEX

Many chapters in the book have emphasized that there are increasingly groups of people who might be regarded as vulnerable, often rendered so by social change and by development conditions. Young children, particularly in poor countries and families, have been a traditional focus of concern, for very good reason. The silent emergency or quiet catastrophe, as the United Nations Children's Fund (UNICEF) calls it, is that 40,000 children die every day, principally from malnutrition and infectious disease, 150 million live in poverty, and 100 million 6–11-year-olds receive no schooling. The catastrophe is exacerbated because the means to avoid many of the deaths are already available from immunization, and it tends to be social and economic conditions that prevent this. Great strides have nevertheless been achieved in the lessening of child mortality from infectious diseases; perhaps less has been done to extend education, and structural adjustment policies are probably worsening this aspect of child well-being.

Today, the health of women is viewed as in need of extension beyond maternal and child health. Chapters 8 and 9 examine this in some detail. Women are at risk not only from the direct child-bearing aspects and the sequelae of child-bearing but also from sexually transmitted diseases, violence and working conditions. Many studies point to young girls, particularly, as being accorded a lower priority in terms of food provision, health care and education than other members of families. While this is not always the case, there is clearly a need for the multiple roles and responsibilities of women, and their particular vulnerability in health and development, to be recognized, and for policies of empowerment to be facilitated.

Increasingly, women are being affected by AIDS as much as are men. In some cities in Africa, the Americas and Europe, AIDS is now the single biggest killer of women aged 20–40. The Panos Institute (1990) points out that many groups of women are at risk from HIV infection, not just prostitutes working in the sex industry. Women are seen to face triple jeopardy from HIV/Aids. A woman may become infected with HIV herself and is likely then to develop AIDS. If she is HIV positive, she may pass the infection to her baby in her womb, who may be born HIV positive. Thirdly, because women almost invariably are the main carers for the sick, they will carry the burden if someone close develops AIDS. The Panos Institute discusses the ways in which women may protect themselves from infection. The conclusion is, again, that this is mainly an issue of relative power in societies. The stronger women's place is in society, the greater are women's options for HIV avoidance.

The other major group who may become vulnerable under conditions of social and economic change are elderly people, as noted above. Chapter 10 focused in particular on the interrelations between socio-economic change and the material, welfare and health situations of elderly people in the poorer countries. Many Third World countries are beginning to recognize the needs of their elderly people. A number of Caribbean countries, many in Asia and some in Latin America are starting to draw up and even to implement social care policies. A generalized response, however, is that 'care of the elderly' is a *family* responsibility and as such it is unnecessary, or even unseemly, for the state to intervene. Such rosy views of the situations of elderly people ignore the fact that many are living miserable existences without any formal financial resources, and are given minimal respect by their families. Single elderly females who never married or who had no children can be in a particularly unenviable condition in many Third World countries. Other elderly people may be left alone in rural areas by children who have migrated to the cities in search of work. In these circumstanes, cash remittances cannot substitute for the assumed family support in countries where formal care systems are missing. Some countries such as China face a particular problem of ageing as a relatively successful one-child policy will render many future elderly people without a child living nearby upon whom to depend.

By contrast, there are many instances of successful policies and provision for elderly persons. Many older people perform valuable functions in the home, as child carers, enabling those of working age to go out to earn, or as sources of advice and wisdom in rural settings. This ideal of 'successful ageing' should not prevent the development of coherent policies and support, however. We must be aware that official support for community care can easily mean reliance on families who do not have the resources and expertise to deal with their elderly members, particularly in conditions of poverty.

Other groups who can be adversely affected by development include handicapped and mentally ill people. Many aspects of physical disability can result from accidents, especially in the industrializing Third World. The incidence of occupational disability is only too obvious in many developed countries, particularly those with a heavy industrial legacy. The social support mechanisms that might have helped such people may be disintegrating as societies develop, and such social consequences need to be

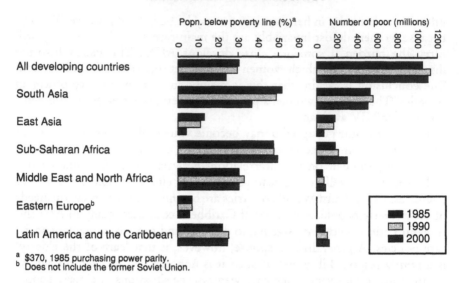

Figure 20.2 Poverty in the developing world, 1985–2000
Source: World Bank (1992)

carefully monitored. Similarly, while there is no doubt only minimal evidence that mentally ill people were well supported in pre-industrial societies, and in many of taday's Third World countries their conditions may be deplorable, this vulnerable group needs particular attention. Neurological and psychiatric conditions are now recognized as perhaps the single major source of disability and morbidity in the world, and, as populations age and as people are exposed to all kinds of stress, there is little to suggest that mental illness will reduce as a key area of socio-medical need.

Underlying many of the above areas of vulnerability is the pervasive factor of poverty. Many of the authors have pointed to relative but often absolute poverty as being the major factor undermining health and the chances for improved health. Poverty can be a consequence or concomitant of many things: unequal distributions of resources, poorly implemented development policies, political conflict, structural adjustment and expenditure cutbacks. These result in under-employment, unemployment and poor public and private services. Undernutrition, malnutrition, poor lifestyles and many other factors interact. Poverty affects individuals, families and entire groups and countries. Sadly, poverty at a macro-scale has increased in the recent past and shows little likelihood of diminishing significantly in the near future (Figure 20.2). Indeed, the proportion of population below the poverty line in sub-Saharan Africa is likely to increase this decade to include about half of the people in the geographical area, meaning that some 300 million people or more will be living in officially defined poverty. Many more will be living at very poor levels if not below an official poverty line.

Table 20.2 Official development assistance (ODA) to the poorest countries

Ten developing countries with highest number of poor	Number of poor (millions)	Poor as % of total world poor	ODA per capita (US$)	ODA as % of total ODA
India	410	34.2	1.8	3.5
China	120	9.9	1.8	4.7
Bangladesh	99	8.3	18.0	4.7
Indonesia	70	5.8	9.3	3.9
Pakistan	37	3.1	8.8	2.5
Philippines	36	3.0	20.3	2.9
Brazil	33	2.8	1.1	0.4
Ethiopia	30	2.5	17.7	2.0
Myanmar (Burma)	17	1.4	4.7	0.4
Thailand	17	1.4	14.1	1.8
Total	869	72.4	4.2	26.8

Source: UNDP (1992:42)

Targets of even the late 1980s and early 1990s of an early reduction of numbers living in poverty seem, by 1993, to be sadly unattainable under realistically expected conditions. Health, environment and economies will as a result continue to suffer.

Many richer nations have attempted to deal with poverty and development needs via aid programmes. Official development assistance has, however, been disappointing in its quantity, equity, distribution and impact. The United Nations Development Programme (UNDP 1992) has identified that much aid is not well targeted. Only about one-quarter of aid goes to the ten countries that have about three-quarters of the world's poorest people (Table 20.2). The 'richest' 40 per cent of the developing world's population receives more than twice as much aid per capita as does the same percentage of the poorest. Redirecting aid in times of financial stringency in the West to areas in Eastern Europe and the former Soviet Union may only mean that the poorest are deprived to benefit the next poorest. Problems of inadequate aid provision are equalled or exceeded by the difficulties of using the assistance that is available effectively.

MEDICINES, DRUGS AND IATROGENIC ILLNESS

Directly or indirectly, a number of authors have identified issues relating to the availability and use of medicines as issues as development occurs. The pharmaceutical industry certainly has an important role to play in the development of medicines and their availability at affordable prices in various countries. However, as profit-oriented entities, many pharmaceutical companies do not see a high priority in developing medicines for mainly poor-country diseases such as malaria. Nor is there much profit in

311

sending the most modern and expensive drugs to countries which cannot afford them. There is thus a rather long and disreputable history of dumping of perhaps out-of-date medicines or drugs that have been withdrawn or used under close supervision in developed countries. The majority of the pharmaceutical industry is well motivated and regulated, but this does not prevent misuse of their products where education, controls or enforcement are deficient.

In many countries, there is a growing recognition of what are sometimes called iatrogenic, or medically caused, diseases. These may be from over-dosing with medicines or from interactions from over-medication, often seen among elderly recipients of multiple drug regimes. There are also undoubtedly many unnecessary invasive surgical investigations made in many health systems, an over-zealous application of laboratory examinations and excessive use of x-rays and CAT scans. These abuses are by no means only found in developed countries. Many residents of middle-income and even the poorest countries are very inclined to turn to inappropriate high-tech examinations and treatments and are, sadly, often encouraged in this for financial reasons by the medical profession. In many countries, if a patient can afford it, it is possible to buy the most advanced radiological examinations, medical and surgical treatments, regardless of whether these are necessary or effective for the supposed condition.

Parallel to the misuse of Western-type medicine and techniques is undoubtedly a vast amount of use of inefficient and inappropriate traditional medicines. As noted in a number of chapters, traditional medicine can prescribe culturally acceptable therapies, and many herbal remedies have pharmacological and/or psychological effectiveness. However, as noted in many developing countries and discussed in relation to Latin America (Chapter 15), there are many quacks and so-called traditional healers who play on local expectations and credibility to ply their wares. While many cures may do little harm, there is growing evidence of harmful side-effects from some traditional cures. Certainly, some culturally determined practices such as female genital mutilation (female circumcision) can endanger the reproductive health and comfort of women and even threaten their lives.

THE MOVEMENT OF POPULATION

Population movements on a sometimes vast scale are a farily common feature of development. In peace, rural–urban drifts bring concentrations of people into towns and cities, to be exposed to new ranges of health and environmental risk. Development projects often involve temporary or longer-term migrations of workers or new settlers who can live in precarious conditions with minimal health provision. In times of war or political turbulence, or as a consequence of natural disasters such as floods, hurricanes and earthquakes, populations frequently migrate for safety. Many people become more or less permanent refugees, as is very common in areas of Africa, South-east Asia, the former Soviet Union and former Yugoslavia, to name but a few areas of refugees in the 1990s. Refugees often have special health needs in excess of those in a fixed population. Their shelter is, by definition, disrupted; their nutritional requirements can sometimes not be met in quantity or quality; and they are often physically or psychologically traumatized from their experiences. The often huge numbers of refugees can be a major drain on the resources of poorer recipient or host countries, leading them to require additional international assistance within their boundaries to care for the citizens of other places.

Population movements can bring people into contact with new or unexpected diseases. Migration has with a fair degree of consistency been identified as a cause of disease diffusion. In the past, cholera has thrived among pilgrims as well as in refugee camps. Erstwhile urban dwellers who have migrated to forest periphery frontiers have been exposed to malaria and dengue in a number of countries such as Brasil, the extent of illness being so great as sometimes to threaten the continuation of rural extension schemes. Migrants for work may come into contact with sexually trans-mitted diseases, and the incidence of HIV infection among mobile workers especially in road transport has been widely recognized in many African countries. Countries which are the recipients of voluntary migration such as Australia often impose stringent health tests and requirements before new migrants are permitted to settle, to avoid the importation of diseases and the financial consequences of ill-health to the nation.

INFECTIOUS DISEASES AND RESURGENCE

Following from the previous point is the emergence of certain new killer diseases such as HIV/AIDS and the resurgence among those with reduced immunities of almost-defeated infections such as tuberculosis (TB). This is common in many cities in the West, notably in New York and San Francisco; tuberculosis is now a major public health hazard which was previously receding in many parts of Africa and Asia. Much of the disease in its resurgent form is resistant to standard drug regimes and even to expensive chemotherapy cocktails.

Similar resurgence has been widely documented in malaria. The earlier success of eradication campaigns ironically caused complacency in a number of places such as India and Sri Lanka, and inefficient continuation of control measures permitted resurgence, as the environmental settings for the disease had not changed. The emergence of resistance to pesticides among the mosquito vectors is well recognized, but many people do not appreciate the widespread resistance to prophylaxis and treatment of many forms of the disease. It is now a major health hazard in a number of areas of the world and one which places serious constraints on development.

313

At times, public health campaigns and social development campaigns can come into conflict. A problem has occurred in Thailand and in some neighbouring countries. In Thailand, there have been concerted efforts to increase the use of condoms by prostitutes in the commercial sex industry. At the same time, there has been a programme to eliminate or reduce child prostitution (in a country where clients are increasingly seeking younger girls in the mistaken belief they might be free from HIV infection). The anti-child-prostitution campaign has led to formerly overt brothels being driven into clandestine operation, and this has meant that the educators attempting to teach safer sex practices and the use of condoms have been unable to reach the prostitutes. As a result of the unintended conflict between these two programmes, the prostitutes have been deprived of health education and legal protection and are probably at greater risk than before from HIV/AIDS.

Increasing attention is being given to the 'hidden' costs of AIDS, and it is being increasingly viewed as a major challenge to development. Throughout the developing world, particularly in sub-Saharan Africa and, potentially in Asia, AIDS is a real developmental factor, limiting the capacity for growth. Co-infection with tuberculosis is also an accelerating problem, with the World Health Organization estimating that 4.4 million people, 98 per cent of whom are in the developing world, are co-infected with HIV and TB. Apart from the direct health care costs and the human misery, HIV/AIDS has two other major characteristics that affect development. It attacks the most economically active sector in any population, those aged between 20 and 45, and it is often most prevalent in areas of poverty, increasing poverty and impairing the ability to initiate economic and social development (Panos Institute 1992).

WAR, CONFLICT AND POLITICAL VIOLENCE

A further theme that has pervaded many of the chapters of the book is that social and political disruption is a major threat to health and health care. This sadly shows few signs of reducing in the near future. Many parts of the world are embroiled in ongoing and bitter wars, often civil wars in effect. Many parts of Southern Africa are prone to direct war injury from mines, bombs and bullets. However, the indirect effects of conflicts in disrupting health care, education and welfare services can be as damaging in the long run. In areas of conflict, it is often very difficult to sustain disease eradication programmes for malaria, for example. It is also very difficult to conduct immunization campaigns such as the childhood Expanded Programme of Immunization discussed in Chapter 9. Community conflict will almost invariably lead to a breakdown of programmes that depend on community participation; and therefore, previously promising health care systems, such as in parts of former Yugoslavia, can be almost destroyed.

The effects of civil war and widespread conflict disrupt health services and at the same time increase the need for curative and rehabilitative care. Famine and malnutrition are often induced in conflicts, as in Somalia in the 1990s. The devastating disruption to many economies accompanying war and political violence can also have very lasting effects on public health, education and related sectors. By diverting economic and human resources and energies into violence and struggle for survival, there is inevitably less to be put into welfare and health activities. The post-violence victims, injured physically and psychologically, rarely receive the care they need in these poor conditions of development.

THE IMPACTS OF DEVELOPMENT POLICIES ON HEALTH

The impacts of development policies on health can be many and varied, direct and indirect. Many impacts are positive, and many programmes for economic and social development have had very positive impacts on health. There are numerous examples of slum improvement and upgrading schemes that have led to improved environmental and public health. However, such schemes can often force out certain groups and individuals, whose health and welfare can be jeopardized as a result (Stephens and Harpham 1991). Other urban expansion projects can, as noted earlier, bring residents into contact with a range of unfamiliar infectious diseases.

There have been many examples of adverse effects of physical development schemes on health. Dams, hydroelectric schemes and irrigation have often in the past led to increases in malaria and diseases such as schistosomiasis (Hunter et al. 1993). While the record of such schemes today has generally improved (in terms of health considerations being taken into account more widely), development schemes in other sectors continue to threaten health (Weil et al. 1990). Industrialization seems to bring increasing direct and indirect threats to health in many industrializing countries, at least until a stage of development is reached when legislation, environmental and health-and-safety controls, as well as public pressure, mean that investment in preventive mechanisms becomes mandatory. Few countries are in the fortunate positions to be able to indulge in such policies, and many workers in industry, agriculture and other sectors are daily at risk from exposure to occupational hazards, toxic substances and unscrupulous workplace practices.

There are numerous possibilities of adverse effects of development strategies forced on to countries by economic circumstances. The widespread damage to health of many macro-economic policies is increasingly being recognized, but few economies are robust enough to follow truly welfare altruistic routes. Perhaps the most pernicious effects are to be seen from the structural adjustment types of policies whose impacts are discussed

315

above and referred to in a large number of places in the book. As noted, such policies, involving retrenchment in public-sector provision, almost always have the most serious effects on the poorest and most vulnerable groups in society. However, increasingly in the developed and the developing world, financial stringency is affecting most groups in society; for example, unemployment, with its increased risks to mental and psychological health, threatens all sectors in the majority of countries. The development policies of some groups of countries such as those in Eastern Europe and the former Soviet Union are resulting in a widespread dismantling of previously umbrella health and welfare systems and are throwing many people on to their own meagre, untried resources.

WESTERN AND TRADITIONAL PATHS TO HEALTH

The final specific issue which we identify in this concluding chapter is that there is evidently neither a single, universally applicable health system nor an agreed view of health as a concept. Most training of medical and paramedical personnel still focuses on a scientific, disease-oriented model, with limited allowance for cultural and socio-economic variations. There is also relatively little real use of indigenous, traditional medical personnel and traditional (mainly herbal) remedies. Some systems, notably that in China, have progressed some way in integration, although full interchange between the Western sector of medicine and the Chinese systems is some way off. Similarly, Chapter 4 discusses the Indian example of integration where some states, such as Tamil Nadu, have made some strides in integration at the primary health care (PHC) level, although it is clear in policy and practice that Western, allopathic medicine is generally regarded as superior. Some countries, such as the Philippines, have had a lengthy policy of incorporating traditional midwives in their health services. It is probably in the field of obstetrics that traditional practitioners (generically termed traditional birth attendants) are most frequently to be found. For around two-thirds of births in developing countries the only source of formal help is a traditional birth attendant. Ironically, obstetrics is one of the most highly technical and medicalized areas of 'non-illness' medicine in most Western countries, which indicates how far apart many areas of the world are in medical services and practice.

It appears certain that there is little possibility this century of extending the full range of modern Western medical facilities, even those of the PHC level, to the majority of people in the poorer countries of the world. In the late 1980s on average in the developing countries, fewer than two-thirds of the population had access to appropriate local health services. This falls to only 46 per cent in the least developed countries and to under 30 per cent in many of the poorest countries, especially those in sub-Saharan Africa. Similarly low percentages had access to safe water, and even fewer had access to sanitation, a service available to only 16 per cent on average in South Asia.

Therefore, to have any hope of extending any form of health care to many groups in the poorest countries, particularly in rural areas, it is essential to utilize the available traditional resources where these have been found to be of some clinical and social value. Many authorities suggest that traditional health care should not be regarded as 'second best' but that for many conditions, including psychiatric disorders, some types of traditional therapies may be very appropriate and effective. In particular, there are very varied perceptions between patients and practitioners as to what medicines are appropriate for specific conditions (Sachs and Tomson 1992). However, sustained and scientifically validated research into traditional medicine's effectiveness is limited. This invariably reduces the acceptance of traditional medicine by the medical establishment in many countries and by health administrators, many of whom are Western-style doctors or come from other formal academic backgrounds, often removed from and suspicious of traditional practices. It is likely that, for each successful example of the use of traditional medicine, the sceptic can find a case of quackery and charlatanry to counter it. As long as this situation continues, it will limit the extent and sincerity of the integration of traditional practitioners into the formal health sector. Nevertheless, traditional practitioners are a resource utilized by vast numbers in many countries, and we must recognize this in policies and future planning.

CONCLUSION

This book has brought together many important issues related to health and development. Its structure has been intended to build on the general issues and concepts introduced in Part I, to show their impacts on certain groups and sectors in Part II and to describe the realities in specific countries and areas in Part III. We have as editors and authors tried to stress the diversity within systems and countries; even within groups such as 'the poor', research is pointing to great diversity of resources, education and opportunity. We have tried to show that there is great variation in the quality of services and care available to people within and between countries. It is essential for future improvements that such issues are taken up by readers of books such as this, who can point to inequalities, show examples of good and bad practice and, perhaps, generally help to raise awareness of what needs to be done and what might be achieved.

REFERENCES

Hunter, J.M., Rey, L., Chu, K.Y., Ade Kolu-John, E.O. and Mott, K.E. (1993) Parasitic Diseases in Water Resources Development: the Need for Intersectoral Negotiation, Geneva: WHO.

Murray, C.J.L. and Chen, L.C. (1992) 'Understanding morbidity change', *Population and Development Review* 18(3): 481–503.

Ory, M.G. and Bond, K. (eds) (1989) *Ageing and Health Care: Social Science and Policy Perspectives*, London: Routledge.

Panos Institute (1990) *Triple Jeopardy: Women and AIDS*, London: Panos Institute.

—— (1992) *The Hidden Cost of AIDS: The Challenge of HIV to Development*, London: Panos Institute.

Phillips, D.R. (ed.) (1992) *Ageing in East and South-East Asia*, London: Edward Arnold.

Sachs, L. and Tomson, G. (1992) 'Medicines and culture: a double perspective on drug utilization in a developing country', *Social Science and Medicine* 34(3): 307–15.

Stephens, C. and Harpham, T. (1991) *Slum Improvement: Health Improvement?*, PHP Departmental Publication No. 1, London: London School of Hygiene and Tropical Medicine.

United Nations (1992) *Demographic Causes and Economic Consequences of Population Aging*, New York: UN Economic Commission for Europe and UN Fund for Population Activities.

United Nations Development Programme (1992) *Human Development Report 1992*, New York: Oxford University Press.

Verhasselt, Y. (1993) 'Geography of health: some trends and perspectives', *Social Science and Medicine* 36(2): 119–23.

Warnes, A.M. (ed.) (1989) *Human Ageing and Later Life: Multidisciplinary Perspectives*, London: Edward Arnold.

World Bank (1992) *World Development Report 1992*, New York: Oxford University Press.

INDEX

floods 43–4
Florida 44
fluorine poisoning 271
folk medicine *see* traditional medicine
food: crisis 200; entitlement 202; environmental change 39–40, 47; security and adjustment 56–7; urban poverty and 115
food poisoning 7, 43
food supplements 141, 148–9
Foot, D.K. 168
Ford, N.J. 90, 92
formal production 133
Foster, H.D. 7
fraud 105
Freire, P. 186
Frenk, J. 12, 14, 15

Gagnon, J.H. 87
Gambia 188
GDP: growth rate 201; per capita 19, 20
Gelfand, M. 226
gender: discrimination 140, 308; employment effects of care-giving 176; utilization of health care services 136; women's health and 124–7, 136
generic drugs 98, 101–2
geography 126–7, 302; women's health 127–36
Ghana 57, 101, 205; cost recovery 60; entrepreneurial health system 211, 213; resurgence of diseases 59
Giri, K. 25
Glantz, M.H. 202, 208
global warming 7, 40–7, 47–8; health effects 41–7, 48, 302; *see also* environmental change
GNP: growth in Latin America 235–6; per capita in LDCs 199, 201, 202–4, 206; Southern Europe 289–90
GOBI-FFF strategies 141–50, 151–2, 196
goitre 271
'Golden Triangle' 92–3
Good, C.M. 66, 67, 70, 71, 72, 79
Goodenough Island 136
governments: and community participation 187–8, 193; and elder care 178; role and multinational pharmaceutical companies 101–2
'gratitude money' 285
Greece 289, 290, 291, 296–8, 298
greenhouse gases 40–1, 47–8; *see also*

global warming, environmental change
growth monitoring 141, 143
Guatemala 235

Haiti 115, 234, 242
halons 34, 36, 40, 48
Hansluwka, H. 17
Hardoy, J.E. 12
Harpham, T. 12, 315; diseases of industrialization 10–12; mental ill-health 116, 117; poverty 10, 115
health: adjustment policies and 22–4, 51, 55–62, 183, 305–6, 315–16; current issues 22–8; definition 3; and development 4–5, 301–2; environmental change and 7–8, 37, 40, 41–7, 302–5; impact of development policies 5–7, 315–16; poverty and 200–2; problems and urban poverty 115–17; scale of problems 24–5; urbanization, modernization and 8–12; Western and traditional paths to 65–70, 316–17
'Health for all by the year 2000' 65, 141, 153, 182, 196, 202, 216
health care expenditures: Africa 252, 253, 257; adjustment and 59–61, 152–3, 183; direct and community services 28; on drugs 105–6; Hungary 282–3; India 219–20; Italy 296; Latin America 243; LDCs 210
health care systems: comparative study 210–13; Latin America 240–2; 'shadow' 285–6
health indicators: Africa 250; China 259–63; Hungary 277–9; LDCs 202–4; middle-income countries 15–20; *see also* infant mortality, life expectancy, mortality rates, under-5 mortality rates
health institutions: China 267–9, 273; India 219, 221, 223, 225, 226; Southern Europe 293–4, 294, 296, 297; Zambia 228–9; *see also* hospitals
health personnel: Africa 252, 253–6; China 269, 273; integration of traditional medicine 73, 75–8, 78–9; Southern Europe 294, 297; trained rural health assistants 232–3; *see also* community health workers, doctors, nurses, village health workers
health planning 15, 116–17, 120, 191, 246

Sanders, D. 58, 59
'sandwich generation' 171–3, 175; *see also* dual care responsibilities
sanitation 117, 200, 237, 303
sanitation and anti-epidemic stations 270
São Paulo 243–4
Satterthwaite, D. 12
Save the Children 193
Scharlach, A.E. 174, 178
schistosomiasis 5, 46, 134, 271–2, 315
Schneider, S. 48
Schram, R. 250, 251
sea-level rise 47, 48
Seers, D. 4, 50
selective-comprehensive debate 151–2, 182–4, 195
selectivity 60–1
self-help 192
Selvin, M. 83
Sen, A. 200, 202
sex industry *see* prostitution
sexual behaviour 85–8, 93–6, 314
sexual interaction: culture and development 87–9, 90; HIV/AIDS transmission 84, 87–9, 90–2, 314
sexually transmitted diseases (STDs) 83–4, 85, 89, 125, 132; *see also* HIV/AIDS
'shadow' health care system 285–6
Shanghai 259, 261
Siddha medicine 69, 72, 74, 219
Sierra Leone 101
SILOS (local health systems) 239–40
Silverstein, J. 93
Simon, W. 87
Simmonds, O.G. 9
Singapore 68, 158; elderly people 158, 162, 180; epidemiological transition 17–19
Sivard, R. 208
skin cancers 37–9, 40, 48
slum improvement projects 24, 115, 118, 192–3, 315
Smith, K.R. 13
Smith, P. 9
snails 271–2
social development campaigns 314
social insurance fund 283
social market economy 287
social marketing 142, 149
social reproduction 134
social security 244
socialist health sytems 211, 212; *see also* Hungary

soil degradation 304
soil erosion 265
solid pollutants 265
Solomon Islands 68, 69
solvent inhalation 237
South Africa 208–9
Southern Europe 289–98; demographic issues 290–1; epidemiological picture 291–8
space, women and 127–9
Spain 278, 289, 290, 291, 294, 298
spatial inequalities *see* inequalities
specialized hospitals 226
squatter settlements 24, 114, 134–5, 192
Sri Lanka 8, 207; comprehensive health system 211, 212–13; drugs 106; life expectancy 19, 206; prices 56; traditional medicine 69
stabilization programmes 53, 54; *see also* structural adjustment
state socialism 279–81, 283–4; *see also* Hungary
Stephens, C. 10, 10–12, 315
Stewart, F. 59
stigmatization, HIV/AIDS and 83–4
Stilkind, J. 9
Stone, L.O. 170
Stone, R. 173
storms 43–4
street children 61
street vending 132
stress, care-givers and 175–6
structural adjustment programmes (SAPs) 53–5, 202; Africa 249, 253–6, 257–8; criticisms of 54–5; education expenditure and 61–2; health care expenditure and 59 61, 152 3, 183, 242–4, 305–6; health impacts 22–4, 51, 55–62, 183, 305–6, 315–16; maternal and child health 152–3; nutrition 55–8; serving the West 54, 63; urbanization and 58–9, 119–20; 'with a human face' 23, 62, 257
sub-Saharan Africa (SSA): adjustment and health 57–8; economic crisis 52–3; education expenditure 61; HIV/AIDS 84–5; migration 58; mortality rates 205, 207; poverty 53, 199, 310; sexual interaction 88; urbanization 113, 114
suicide 237, 244, 281
sulphur dioxide 303